The Living End

The Living End
The Future of Death, Aging and Immortality

Guy Brown

Macmillan
London   New York   Melbourne   Hong Kong

First published 2008 by
Macmillan

Houndmills, Basingstoke, Hampshire RG21 6XS and
175 Fifth Avenue, New York, N.Y. 10010
Companies and representatives throughout the world

ISBN 978-1-349-95369-1

This book is printed on paper suitable for recycling and made from fully managed and sustained forest sources. Logging, pulping and manufacturing processes are expected to conform to the environmental regulations of the country of origin.

A catalogue record for this book is available from the British Library.

A catalog record for this book is available from the Library of Congress.

10   9   8   7   6   5   4   3   2
17   16   15   14   13   12   11   10   09   08

Contents

Contents

For my father

Special thanks to: Sarah Radcliffe, Sara Abdulla, Christoph Goemans, Vilma Borutaite, Mike Murphy, Donald Smith, and Mervyn Singer

1
beginnings and endings

My grandmother leaned forward and whispered conspiratori-
ally: 'I have been wanting death for the last five years'. It
seemed like she was letting me into a profound secret, but I
knew she had been telling the same to anyone who would
listen for at least the last ten years. Still, it seemed an intimate
kind of confession, almost a dirty secret – not the kind of thing
you should be confiding to your grandson. I didn't want to
know about her relationship with her own death; that was private – something to be faced alone amongst the terrors of the
night. But her confession also had an edge of fearless com-
plaint – damning the world and her allotted fate of suffering
and boredom. She wanted to embrace death rather than run,
hide and slowly diminish in terror. At the same time she was
admitting her unbearable powerlessness: she wanted death,
but even this ultimate surrender was not accepted – death was
playing hard to get. This brief encounter with another's death
made me think about how one could come to desire one's
own death, either as an escape from life or as a positive end in
itself. Were death and old age something other than I had
imagined?

My grandmother is long dead now; and I am myself voyaging through middle age. In the last few years my mother-in-
law has become demented, my father has died of cancer, and I
have acquired a degenerative disease. My own death looms
on the horizon. But when I was young, I saw things differently,
I was determined to confront death and seek out immortality.
That quest could have taken me in many different directions,
such as religion or the quest for fame, but being literal minded
I was drawn to science.

During the day, I am a research scientist working on the
molecular mechanisms of cell death and degenerative disease
at the University of Cambridge. And in this book I describe
how the new sciences have changed our understanding of
death, and how death is likely to change in the future. But

during the night, I lie awake and contemplate my own death. And in writing this book I am also setting out to understand how the revolutionary changes in death and the life sciences will impact on our concepts of life and death.

Once upon a time death itself was young and virile; remorselessly scything through humanity; pervasive, undeniable and unstoppable. Now death has grown old, and is slowly falling apart; a suppressed terror consigned to some attic of our memory. Some say that God died in the nineteenth century and our humanity in the twentieth century. Will death itself finally expire in the twenty-first century, holding out the prospect of immortality, and with it the possibility of us humans becoming gods?

Death is certainly not what it was. Life in the past was once described as 'nasty, brutish and short', but this would be a better description of death throughout most of history. The very shortness of life tended to mean that death too was short. People died either as children or in their prime, so aging and aged individuals were rare. The most common causes of death were infections, violence and childbirth. On the whole, death was rapid: people were fully alive one day and fully dead the next. There was relatively little grey area between life and death. Of course there were exceptions: people suffered from disabilities; some lucky individuals lived long enough to age; some people died from long lingering diseases. But the average death was shorter in the past and occurred in a much younger individual.

In the developed world, and increasingly in the developing world too, we no longer die of infectious diseases, we no longer die quickly, and we no longer die young or in our prime. Our concept of death, inherited from history and film, is either of violent death or of a fevered soul wracked by delirium fading into the night and expiring with a final, faint breath. This might have been the average death two hundred or four thousand years ago, but the reality now is very different.

Today, the average lifespan in the developed world is about 78 years. And this lifespan is increasing by two years every decade. That is, every decade you live, your expected age of death recedes by at least two years, or 5 hours per day, or 12 seconds per minute. If this trend persists (as it has done for at least 100 years) someone born today would be expected to live to the age of 100 years.

Sickeningly, there remain regions of the world still ravaged by acute causes of death, such as AIDS, starvation and war in parts of Africa. But as a whole, the developing world achieved more dramatic increases in lifespan and decreases in acute forms of death during the twentieth century than in the developed world. Even in the developing world, people are now living long enough to mainly die from degenerative diseases.

Our concepts of 'old' and 'old age' are out of date. It is no longer useful to categorize everyone over the age of 65 as just 'old' – there are the 'old' (65–85 years), the 'very old' (85–100 years) and the 'extremely old' (over 100 years). Just because we are old, or know people who are old, does not mean we know what it is like to be very old or extremely old. These are radically different phases of life, as different as youth from middle age, or middle age from old age.

We cannot assume that what applies to old age equally applies to extreme old age. We know very little about this phase of life, but this is the fastest growing fraction of the population. If current trends persist the UK Government Actuary's Department predicts that the UK population over 65 years of age will triple from 4.6 million now to 15.5 million in 2074, and the population over 100 years of age will increase 100-fold from 10,000 now to 1 million in 2074. The future is not just old, or even very old, but is also extremely old. We are voyaging into a new realm of human life that has hardly existed before, and about which we know very little.

While we have been remarkably successful at delaying death, crucially, we have failed to delay aging. An 80-year-old person today appears to be just as aged as an 80-year-old several hundred years ago. This has huge consequences for life and death, both now and in the future. The incidence of degenerative diseases, such cancer, Alzheimer's disease and cardiovascular disease increases dramatically with age. So as the population inexorably ages, these maladies that were formerly rare to nonexistent become commonplace. A plethora of frightening new neurodegenerative conditions, such as Pick's disease, Lewy body disease and fronto-temporal dementia, have apparently appeared out of nowhere as life expectancy marches ever higher.

Currently in the US 46% of people over the age of 85 are thought to have Alzheimer's. If current trends persist, someone born today has a one in three chance of dying with dementia! This is not just a problem for somebody else in the future, unfortunately: it is a problem for you and me now. Current life expectancy (without any extrapolation) for someone aged 65 years in the UK is a further 17 years for a man and 20 years for a woman. So a 65-year-old woman now can expect to live to 85 years (and extrapolated life expectancy could take that to 90 years), so that the probability of developing dementia is about one in four. If we include Mild Cognitive Impairment (MCI, a precursor of Alzheimer's), then two thirds of 65-year-old women would be expected to develop MCI or Alzheimer's before they die.

Imagine the consequences if this nightmare becomes a reality. Imagine the psychological consequences of expecting to die with dementia. Imagine the economic consequences of providing round-the-clock, one-to-one care for millions of demented or disabled people for decades. Imagine the social consequences of loss of faith, loss of hope and loss of trust.

And what are we doing about it? Virtually nothing. Death, dying and dementia are nowhere on the political agenda. We are too afraid to think about the 3 Ds. And that suits the politicians fine – because of course it wouldn't be easy to do anything about them.

Even now the vast majority of people in the developed world (and increasingly in the developing world) die from degenerative diseases, such as cancer and heart disease. These diseases are caused by age, and dying from them is slow and is becoming slower, so that the processes of death and aging are merging into one. Death is currently preceded by an average of 10 years of chronic ill health, and this figure is rising. But aging starts much earlier. Many of our physical and mental capacities peak at around 20 years of age and then undergo a long, slow decline. Few people survive until death without significant physical and/or mental disabilities, extending over decades. Death is no longer an event, it has become a long, drawn-out process.

The word 'death' has in the past had at least three related but distinct meanings. Firstly, it refers to the process of dying, as in 'The Death of Ivan Ilitch'. Secondly, it means the terminal event at the end of life and dying, as in 'he finally expired and gave up the ghost'. Thirdly it denotes the state of being dead, as in 'he is dead'. These three meanings are located before, at and after the terminal event of life. However, if the dying process and the terminal event itself have become stretched over many years, then the three meanings of 'death' become mixed up, with themselves and with aging and degeneration. We need a new theory of death.

The digital theory of death is dying. We can no longer think of ourselves as suddenly going from being fully alive (1) to fully dead (0), in the same way that we have accepted that we do not jump from being non-existent (0) to fully alive (1) at birth. Becoming a full human being is a process. We grow into it over a period of years and decades, but then we grow out of it. That does not mean that aging is growing up in reverse. They are obviously completely different processes: an extremely old person is not the same mentally or physically as a new-born baby. But growing up and growing old could be thought of as growing into and growing out of life. This would give us an analogue theory of life and death: there is a continuum between life and death. Life is not all-or-nothing – there are degrees of life, and at some times in our lives we are more alive than at other times. We would all agree that at some times of the day (or night) we are more alive than at others. But to assert that some people are more alive than others is a political bombshell. However, whether we like it or not, the future reality, where the majority of us die demented or cognitively impaired, will force new concepts of life and death upon us.

> Grave men, near death, who see with blinding sight
> Blind eyes could blaze like meteors and be gay,
> Rage, rage against the dying of the light.
> And you, my father, there on the sad height,
> Curse, bless, me now with your fierce tears, I pray.
> Do not go gentle into that good night.
> Rage, rage against the dying of the light.[1]

In Dylan Thomas's visceral poem, life is light and death is darkness; and in between there is the enveloping twilight of old age and aging: 'the dying of the light'. In this metaphor, death, dying and aging roll into one continuous process: a dimmer-switch theory of death and aging! A dimmer-switch is an analogue device, enabling a continuously variable amount of light. We should contrast this to the digital switch theory of death – a switch that has only two positions: on or off, corresponding to fully alive or fully dead. Thomas gives us a richer

theory whereby death is mixed in with life as shades of darkness and light that may grow and fade with day and night.

But the dimmer switch gives us a rather uni-dimensional,
monotonic metaphor for life and death. We need some colour. Death is not just happening later and taking longer, it is
also fragmenting. There are different types of death occurring
in the same person at different rates and to different extents in
different people. There is death at different levels: molecular
death, cell death and organ death; death of the individual,
death of the culture and death of the species. There are multiple deaths of different parts of our body and mind: death of
our physical abilities and appearance, death of our various
mental capacities. There is reproductive death, social death,
and psychological death. There is death of desire, there is the
death of memory, there is death of the will to live. All these
things fade away at different ages, at different rates, and to
different extents.

Death is no longer a unified event. It is shattered into multiple uncoordinated processes. This is one reason it is impossible to quantify life or death. If someone's brain is 100% dead
and body 100% alive, can we say they are 50% dead and 50%
alive? I think not. But neither can we say they are fully alive or
dead. Nor can we hide in such simplicities when Alzheimer's
disease has gnawed away half of someone's brain.

We as a global society have been remarkably successful at
taming acute forms of death. Until as recently as 100 years
ago, acute death was the norm – now it seems like an outrage.
Yet many acute forms of death have been converted to chronic death or disability. Heart attacks have become heart failure;
stroke has become vascular dementia; and diabetes, AIDS,
and even some cancers have been converted from acute
causes of death to chronic disabilities. All of these are great
medical advances, but they have a downside: the conversion
of acute to chronic death. This has an upside for the pharma-

ceutical companies that now dominate medical research. Curing diseases does not pay – because you lose your patient – whereas converting an acute disease into a chronic disease pays very handsomely indeed – because you convert a short-term patient into a long-term consumer of your drugs.

Patients and medics have, perhaps understandably, been more concerned to prevent death than to prevent disease. Sudden death is generally more 'sexy' than chronic death. Just compare the column inches devoted to the acute death of young people (e.g. Princess Diana's car crash) compared with coverage of the chronic death of old people. Yet the latter death is much, much harder for the individual concerned. We may think that chronic death from old age is 'natural', while acute death of the young is somehow unnatural. Of course, the exact opposite is the case. Death from old age is extremely rare in wild animals, and was rare in humans until one hundred years ago. Only in the unnatural conditions of modern society and medicine can the exotic diseases and deaths of old age bloom. Many of these diseases and causes of death have only been recognized in the past few decades, and many more undoubtedly lie undiscovered ahead.

In Greek mythology, Tithonus was a handsome mortal who fell in love with Eos, the goddess of the dawn. Eos realized that her beloved Tithonus was destined to age and die. She begged Zeus to grant her lover immortal life. Zeus was a jealous god, prone to acts of deception in order to seduce beautiful gods and mortals, and he was not pleased with Eos's infatuation with a rival. In a classic Devil's Bargain, he granted Eos's wish – literally. He made Tithonus immortal, but did not grant him eternal youth (which Eos and the other gods obviously had). As Tithonus aged, he became increasingly debilitated and demented, eventually driving Eos to distraction with his constant babbling. In despair, she turned Tithonus into a grasshopper. In Greek mythology, the grasshopper is immortal, and this myth apparently explains why grasshoppers chirrup ceaselessly, like demented old men.

Tithonus's fate now threatens us all – ever-increasing lifespan, but with ever-increasing enfeeblement, ill health and dementia. This is the Tithonus scenario, and we as a society need to decide whether we accept this fate or should do something about it.

Throughout history people have sought to escape death and claim immortality in three different ways: spiritual, genetic and cultural. We may survive as spirits in some afterlife. We may survive through our genes passed on to our children and children's children. We may survive though our works, deeds and memories – 'memes' – left to our family, friends and society in general. The desire to survive spiritually, genetically or culturally has been, and continues to be, one of the most important motivators in life. Indeed, society as we know it (religion, family, culture) would be impossible without such motivators.

Whether or not we believe that a spirit survives after death, genes and memes certainly can. But they too can die or slowly fade away. Our children's children may not survive or breed. Even if they do, our genes will be slowly diluted and mutated beyond recognition. More importantly perhaps, an individual's life can leave a lasting impression on society and on those who knew him/her, via the memories and surviving effects of the deceased's ideas, works and deeds. Such things might be as small as a remembered kindness, or as large as a scientific theory. In the end, all of them, in different ways and rates, will fade and die. Prior to what we call the terminal event of death, little bits of life fade away (our abilities, energies and desires), but the process continues after death, as what is old is inexorably replaced by what is new.

The analogue nature of death has important implication for our concepts of life and self. Since at least the Renaissance the concept of the individual self has become increasingly central and monolithic in Western thought. The self became the unified, indestructible, unchanging atom of society. Just as we had an atomic theory of matter, the concept of soul gave us an atomic theory of self: the core of a person that was completely separate although similar in different people, unsullied by life, and unanalysable into component parts. But if we accept that death is a process changing multiple components of ourselves throughout life, then we also need to consider the possibility that the self is not the same self throughout life. Are you really the same person you were as a child, or you were ten years ago, or will be at 70 or 100?

The notion that someone may be partly dead seems absurd, until we encounter someone who has 'unsuccessfully' aged or has Alzheimer's. If we entertain the notion that beyond death various components of ourselves, spiritual, genetic or cultural, will survive in non-individual form, then our concept of the monolithic self may fracture and dissolve into a web of interacting components, the genes and memes that we share with our family and culture. The classical atomic theory of self (where the self was single, separate, unified, digital and unchanging) needs to be replaced by a wave theory (where the self is multiple, overlapping, distributed, analogue and changeable). If the self is not digital (all or nothing), this implies that we can lose parts of the self (for example by forgetting or by changing part of the body or mind), but it also suggests that parts of the self may survive the death of the individual.

In the twentieth century death replaced sex as the taboo subject we could not talk about, yet we all end up 'doing it'. The suppressed dread of death has allowed our society to sleepwalk into a situation where people face real horrors at the end of life, simply because we cannot face dealing with the issue of how people should exit life. Death has been banished to hospitals, the worst possible place to end life. Medicine has become devoted to keeping people alive at any costs, rather than helping people die. Huge resources are devoted to preventing infectious diseases and heart attacks, possibly the ideal way to die, which inevitably condemns people to die by more protracted means.

Only by recognizing that death is part of life, and that many people experience a living death at the end of life, can we make sensible decisions about whether people should be allowed to choose a dignified exit from life. We all recognize that we need to make provision for a pension, but how many of us are making provision for dementia? If society really cared about the last ten years of life to the same degree as the first ten years of life, then we would have a real chance of preventing aging and dementia before it was too late. If not, we have the real prospect of creating Hell on Earth, and locating it at the end of life.

Death and aging will be the defining problems of the twenty-first century. If death continues to recede while aging continues, then our world will turn grey and then white. The most obvious impact will be economic: supporting a huge and ever-growing population of increasingly disabled and demented very old people will be crippling to all our economies. The projected scale of the looming pensions crisis still takes a very optimistic view of the balance between death and aging in the future. A more realistic view would see real damage to the economies of both the developed and developing worlds.

Perhaps more seriously, each individual will have to face up to a slow, grinding, degrading journey into darkness, and possibly dementia. The potential of this to change our outlook on life, and the consequent impact on society, should not be underestimated. Global terrorism may directly affect tens of millions of people, and global warming is likely to affect hundreds of millions. But the Tithonus scenario will strike billions in the most direct and personal way possible, rotting their bodies and brains from within.

However, looking on the bright side, many researchers are now seriously suggesting that aging may be completely solved during the twenty-first century. The promise of stem cells, cloning, RNAi and gene therapy, the sequencing of the human genome, and the deep understanding of our bio-chemistry, suggest that it is just a matter of time before we can make humans, for all practical purposes, immortal. However, not everyone is happy at the prospect of universal immortality. Bioethicists and politicians are reaching out to restrain these sciences, for example banning cloning, stem cell research and genetic engineering. But aging and dementia are not natural, and we should not accept them as our inevitable fate – allowing millions to suffer in silence.

A flood of new evidence is starting to have an impact on our concepts of life and death. Genetics and genomics have found multiple genetic switches that can dramatically extend or shorten lifespan (at least in worms!). Evolutionary biologists have given us an understanding of the evolutionary origins of aging. Cell biology has uncovered a programmed process of death that occurs throughout life, and is indeed essential to it. Medical research has made massive progress in understanding how we die. Medicine has confronted issues such as the defini-tion of death, an acceptable quality of life, rationing of health care to the elderly, euthanasia, the treatment of the incurable dying, and the handling of the dead body. The increasing interaction of psychology, philosophy and neuroscience has brought new insights into the composition of self and its changes through life and death. The relatively new science of aging has discovered an inexorable process of decline at the heart of our molecular machinery, but also claims to have found potential ways to interfere with that decline.

Leading aging researchers are now urgently calling for society to prepare for an era of dramatically extended lifespan. Scientists, doctors, economists, social planners, philosophers, theologians and even politicians are waking up to the massive impact that aging and increased lifespan will have on our society. For ourselves and our children we now urgently need to consider the future of death, dying and dementia. And in this book I intend to 'rage, rage against the dying of the light', as well as exploring options for changing our fate.

interlude 1
a brief history of death and damnation

THROUGH ME THE ROAD TO THE CITY OF DESOLATION,
THROUGH ME THE ROAD TO SORROWS ETERNAL,
THROUGH ME THE ROAD AMONG THE LOST CREATION
...
LAY DOWN ALL HOPE, YOU THAT ENTER IN BY ME.

This is a rough translation of the inscription above the gates of
Hell in Dante's *Inferno*. The *Inferno* was the first instalment of
Dante's *Divine Comedy*, written in Italy between 1308 and
Dante's death in 1321, and considered the last great work of
literature of the Middle Ages, and the first of the Renaissance.
After tremulously passing through the gates of Hell, Dante has
himself guided by his mentor, the Roman poet Virgil, through
the nine levels of Hell. First they must traverse Acheron, the
accursed river defining the outer boundary of Hell, across
which Charon ferries the damned souls:

And suddenly coming towards us in a boat,
a man of years whose ancient hair was white
shouted at us, 'Woe to you, perverted souls!
Give up all hope of seeing Heaven:
I come to lead you to the other shore,
Into eternal darkness, ice and fire.'[1]

The journey through the Inferno was supposedly a metaphor
for a spiritual journey, consisting of the stripping away of the self
and its sins, prior to the eventual rebirth of the soul in Paradise.
Analogously, in this book, I will strip away our old concepts of
death and self in nine chapters. Hopefully, at the end, a new concept of mortality and immortality will be reborn from the ashes
of the old. First we must pass through what may at times seem
the rather harrowing territory of death, dying and dementia.
  Because we are embarked on a potentially heavy and
depressing subject, I have sought to lighten the journey with

various distracting interludes along the way. Those of you with more robust constitutions may choose to ignore these frivolities. In this first interlude we will take a whirlwind tour through the history of death.

Death obviously has a long prehistory, but we will pass over the first two billion years or so. Death also knows no cultural boundaries, but we will restrict our attention to the Western concept of death, starting with the Egyptians. Egyptian society, it has been said, consisted of the dead, the gods and the living. Throughout their long history, the ancient Egyptians seem to have spent much of their time thinking of death and making provisions for the afterlife. The vast size, awe-inspiring character, and ubiquity of their funerary monuments bear testimony to this obsession with death. The Egyptians were a practical and literal-minded people, who would have no truck with the idea of disembodied spirits, so their afterlife required a real body, and hence the need to preserve the body after death, as a mummy.

One of the great myths of ancient Egypt, which influenced subsequent concepts of death, was that of the god Osiris. Osiris's evil brother, Set, fooled Osiris into getting into a coffin, which he then shut, had sealed with lead and threw into the Nile. Osiris's wife, Isis, searched for his remains until she finally found him embedded in a tree trunk, which was holding up the roof of a palace. She managed to remove the coffin and open it, but Osiris was already dead. She used a spell she had learned from her father and brought him back to life so he could impregnate her. After this brief interlude, he died again, so she hid his body in the desert. As a consequence of their dalliance, months later, Isis gave birth to the falcon-headed god Horus. While she was raising him, Set had been out hunting, and one night he came across the body of Osiris. Enraged, he tore the body into 14 pieces and again threw them into the Nile. Isis gathered up all the parts of the body and bandaged them together for a proper burial. The Gods were impressed by the devotion of Isis and restored Osiris to 'life' as the god of the underworld. This myth was perpetuated by the cult of Osiris, which lasted for 2,000 years and spawned many other mystery cults.

Three ideas prevailed in ancient Egypt that came to exert great influence on the concept of death in other cultures. The first was the notion of a dying and rising saviour god, epitomized by the myth of Osiris, who could confer on devotees the gift of immortality – this gift was first sought by the pharaohs and then by millions of ordinary people. The second influential concept was of a post-mortem judgment, in which the quality of the deceased's life would influence his ultimate fate. The third idea, which may be a very old idea indeed, was that the newly dead had to make a journey from the world of the living to the underworld or over-world of the dead. This perilous journey has at various times involved tackling magical gates, lakes of fire, terrible monsters and a sinister ferryman.

To assist that grim journey, various guides have been provided. From about 1500 BC, Egyptian tombs were helpfully stocked with the Book of the Dead, containing spells for dealing with perils encountered *en route*. Orphic communities in Italy gave directions for the next world on gold plates deposited in the graves. Mediæval Christians were provided with advice about dying in a book called *The Art of Dying*, while Tibetan Buddhists used the Tibetan version of the *Book of the Dead*. Chinese Buddhists had a graphic guide to the ten hells of their next incarnation in the form of popular prints. This book you are reading is intended to provide a more up-to-date version of an armchair guide to death and dying.

Ancient Mesopotamian literature records the visit of the goddess Ishtar to the realm of the dead, the way to which was barred by several gates. At each gate the goddess was deprived of some article of her attire, so that she was naked when she finally came before Ereshkigal, the queen of the underworld. This successive stripping of the celestial goddess might symbolize the stripping away of the attributes of life that the dead experienced as they descended into the 'land of no return'.

In much of the ancient world, including Mesopotamia and Greece, all of the dead, both the good and bad, the rich and poor, entered a twilight underworld as shadows of their former selves. Homer's Odysseus visits this world, known as Hades to the Ancient Greeks, and returns to tell the tale:

> When I had finished my prayers and invocations to the communities of the dead, I took the sheep and cut their throats over the trench so that the dark blood poured in. And now the souls of the dead who had gone below came swarming up from Erebus.... From this multitude of souls as they fluttered too and fro by the trench, there came a moaning that was horrible to hear.[2]

Homer here reflects a general fear and dread of the dead, which remains today in the concept of ghosts. Rituals were performed on and for the newly dead to aid their transfer to another realm and prevent them returning to haunt the living. Practical equipment for the journey to the next world was provided for the Greek and Roman dead, in the form of money (a coin in the mouth) to pay the ferryman Charon for their passage across the Styx, and honey cakes for Cerberus, the fearsome dog that guarded the entrance to Hades. Charon was the offspring of Erebus (Darkness) and Nyx (Night), who also begat Death, War and Famine, the dark children of Hades. In later mythology, Charon personified Death himself.

In Hellenistic and Roman times the underworld included subcompartments: Elysium, a paradise for the good, and Tartarus, a hell for the bad. In the Aeneid, written in Rome from 29–19 BC, the author Virgil has his hero Anaeas guided through the underworld by a priestess who tells him:

> 'On the left is the road of punishment for evil-doers, lead-ing to Tartarus, the place of the damned.' ... They could hear the groans from the city, the cruel crack of the lash, the dragging and clanking of iron chains.... 'This is the road we take for Elysium.' Here a broader sky clothes the plains in glowing light, and the spirits have their own sun and their own stars.[3]

In Judaism death was originally the end: 'For the living know that they will die, but the dead know nothing, and they have no more reward' (Eccles. 9:5). There may, however, be an underworld, Sheol, 'the land of gloom and deep darkness' (Job 10:21). In Sheol, the good and the wicked shared a common fate, much as they had in the underworld of Homer, but that fate was undefined, perhaps because nothing much happened there. The concept of a resurrection of the dead originated later, during Judaism's Hellenistic period (4th century BC–2nd century AD). Isaiah announced that the 'dead shall live, their bodies shall rise', and the 'dwellers in the dust' would be enjoined to 'awake and sing' (Isa. 26:19). Both the good and the wicked would be resurrected. According to their deserts, some would be granted 'everlasting life', others consigned to an existence of 'shame and everlasting contempt' (Dan. 12:2). As in Virgil's underworld, Sheol itself became split into three divisions, to which the dead would be assigned according to their moral desserts. The real valley of Hinnom, where Jerusa-lem's municipal rubbish was burnt, was transmuted into a vast camp for the dead, designed for torturing the wicked by fire. This was a precursor of the Christian and Islamic versions of hell.

It was into this world that Jesus Christ was probably born. His untimely death and failure to return as the Messiah exercised later Christian theologians. Delay in the promised Second Coming of Christ led to an increasing preoccupation with what happened to the dead as they awaited the resurrection and the Last Judgment. One view was that there would be an immediate individual judgment and that instant justice would follow: the deceased would be despatched forthwith to hell or paradise. This notion demeaned the impact of the prophecy of a collective mass resurrection, followed by a public mass trial. Moreover, it deprived the dead of any chance of a post-mortem expiation of their misdeeds. The Roman Catholic notion of purgatory sought to resolve the latter problem; regulated torture would expiate some of the sins of those not totally beyond redemption.

John Milton's epic poem *Paradise Lost* (written around 1660 in the aftermath of the English Civil War) gives the Christian account of how death and damnation entered the world with Adam and Eve's disobedience to God, and the resulting inheritance by all subsequent generations of mortality and original sin. Milton's tale begins with Satan and the rebel angels defeated in Heaven and fallen into Hell. But Satan is not contained for long – he builds a new empire in Hell, and there hears a rumour of a new world (separate from Heaven and Hell) created by God and populated by a new being in God's image – he determines to corrupt this new creation in God's eyes, and annex it to his own. Satan gets through to the new world, and, disguised as a snake, commits the dastardly deed in the Garden of Eden: convincing Eve to disobey God by eating of the fruit of the Tree of the Knowledge of Good and Evil. Eve, in turn tempts Adam with the fruit; they promptly have sex, and fall into blissful sleep. God is not amused, and sends the Archangel Michael to inform Adam of his error:

Adam, now ope thine eyes, and first behold
Th' effects which thy original crime hath wrought

In some to spring from thee, who never touched
Th' excepted tree, nor with the snake conspired,
Nor sinned thy sin, yet from that sin derive
Corruption to bring forth more violent deeds.[4]

Thus according to Christianity, death, disease and aging are
the wages of sin, which we have inherited as punishment for
the original sin.

According to Islam (Arabic for 'submission'), God (not Man)
determines everything. God takes away people's souls both
'upon their death' and 'during their sleep'. He 'retains those
against whom he has decreed death, but returns the others to
their bodies for an appointed term' (39:42–43). Death is
repeatedly compared with sleep, which is at times described
as 'the little death'. The Qur'an states: 'Some will die early,
while others are made to live to a miserable old age, when all
that they once knew they shall know no more' (22:5). Muslims
respect dead bodies, which have to be disposed of very
promptly. Cremation and dissection are viewed with abhor-
rence. In part this is because the dead are rapidly visited by the
Angel of Death and must face a doctrinal test within the grave,
followed by Final Judgement and dispatch to either Hell or al-
jannah (the Garden, 'paradise').

Hindus and Buddhists have a rather different attitude to
death, as they believe in physical rebirth/reincarnation in the
same world. They are not particularly happy about this
rebirth, however, as they think that moral failures in one life
are paid for in the next. So they hope eventually to escape
from the cycle of rebirths, and merge with the cosmos. On the
other hand, Hindus think that death is very 'polluting' for the
relatives and friends left behind. Interestingly, this idea of pollution may affect modern Western ideas of death.

One may be tempted to think that Western attitudes to
death have changed little in the past thousand years. How-

ever, French historians of ideas, in particular Philippe Ariès, have uncovered seismic changes in people's views of death over the past millennium. Ariès identified at least three distinct periods or attitudes: mediæval acceptance, renaissance anxiety, and modern rejection of death.

During the Middle Ages the predominant attitude was one of naturalistic acceptance or resignation. Death was everywhere and for everyone to see, almost every day, so it was accepted as a fact of life. It was a transition, just like birth, baptism and marriage, that was expected to lead eventually to eternal life in heaven. And like other transitions, it was ritualized within a social context, rather than being hidden away. Family and friends attended the dying, usually in their own home. The ritual was simple and there was no excessive emotion or mourning. After death the body was entrusted to the Church to await an assured Resurrection at the Second Coming of Christ. The vast majority did not receive an individual burial, tombstone or any lasting memorial to preserve their name. Once the corpse had rotted away in temporary ditches, the bones were either dumped in collective charnel houses or surfaced as so much flotsam in the world of the living. Once dead, the dead were largely forgotten. Their spouses rapidly remarried, their mothers produced new children, their employers found replacement labour.

The Renaissance was characterized by the emergence, or rather re-emergence, of the 'individual' from a thousand years of collectivism. This mirrored the emergence of an atomic/digital idea of self from the previous collectivist view of self. And if our selves are all-or-nothing and we are separate from the world and other people, then of course it matters immensely what happens to us as individuals at death. Images or lives of individuals became legitimate subjects for artists and authors to represent or preserve, replacing the unchanging representation of God and his close family. The deceased's name started to appear on tombstones or memorial plaques. The concept of the Last Judgement emerged, whereby the individual was judged at the end of the world, and thence consigned to either heaven or hell. Subsequently an intermediate location – Purgatory – was invented, and the idea that the living could shorten the dead soul's sentence in Purgatory by praying for the soul or paying the church for an indulgence. Thus the idea of the individual's life and death slowly developed to become a more central concern. After all, if an individual's life mattered, then the cessation of that life mattered too. And a certain anxiety developed about what might come afterwards. Dante's *Divine Comedy* reflects this new concern for the individual's fate after death.

It was not however until the invention of Romanticism and Sentiment in the eighteenth and nineteenth centuries that death became a calamity. Funerals ballooned, emotions overflowed, and mourning went out of control. Huge new cemeteries and death memorials were built in Catholic Europe; personal and family rituals of visiting the grave were invented; festivals of the dead, such as All Saints Day, blossomed. These were all manifestations of a new concern with other people's death.

The twentieth century saw a frightening revolution in attitudes to death. Starting in the USA and Protestant Europe, death became hidden, suppressed and taboo. The real experience of other people's death or dead bodies was completely removed from sight. People could lead their whole lives without seeing a dead body. Death and dying were medicalized and professionalized. Once anyone showed serious signs of dying they were shipped off to a hospital, which meant that the domestic rituals of death could not be performed, and people could not be acquainted with death and dying. But the medical profession was not interested in death itself, but rather with preventing it, which made hospitals the worst possible place to die. Death was denied, there was a fear of telling people that they were dying, or of even discussing death. The doctors, the relatives, the friends and the dying themselves did not want to talk about death. Once dead the body was dealt with by professional body disposers, so that the family need never come into contact with the body. Confrontations with death became so rare and exotic as to seem almost unnatural and polluting: fear of death was endemic.

The cult of youth that developed in the second half of the twentieth century was associated with a fear of aging and frantic efforts to suppress its manifestations. Fading and greying hair was replaced or coloured; sagging skin, breasts and penises were uplifted. Images of youth were everywhere; the reality of age was banished from the glossy magazines and TV screens, and kept safely at home or in 'homes'. Individuals within an increasingly secularized society began to lose faith in an afterlife, and to face the prospect of total annihilation at death. But outside religious faith there was no social framework or discourse for dealing with death.

The fear of death and aging that developed in the twentieth century may seem paradoxical when we consider the spectacular success in beating back death that occurred over the same period. But as we shall see in the next chapter, these two developments are intimately connected. During the late nineteenth and twentieth centuries the average age of death jumped from 35 to 65 years. Consequently, instead of a few people aging, almost everyone could expect to age, and to die from a long drawn out degenerative disease. No wonder people became afraid of death.

2

the changing face of death

Death in retreat

One of the greatest achievements of human history has been the spectacular increase in life expectancy that occurred during the nineteenth and twentieth centuries (Figure 2.1). In 1800, global average life expectancy at birth was about 30 years; by 2000 it had increased to 67 years. That is, the average lifespan for the whole world more than doubled. The most common age of death jumped from infancy to old age (Figure 2.2). Whereas previously people died chiefly from infectious diseases with a short course, now people die mainly from chronic diseases with a protracted time course (Figure 2.3). The numbers of people living in their economically productive years filled out, and the old became commonplace everywhere. Prior to this longevity revolution, reaching old age was rare; today it is the fate of almost everyone.

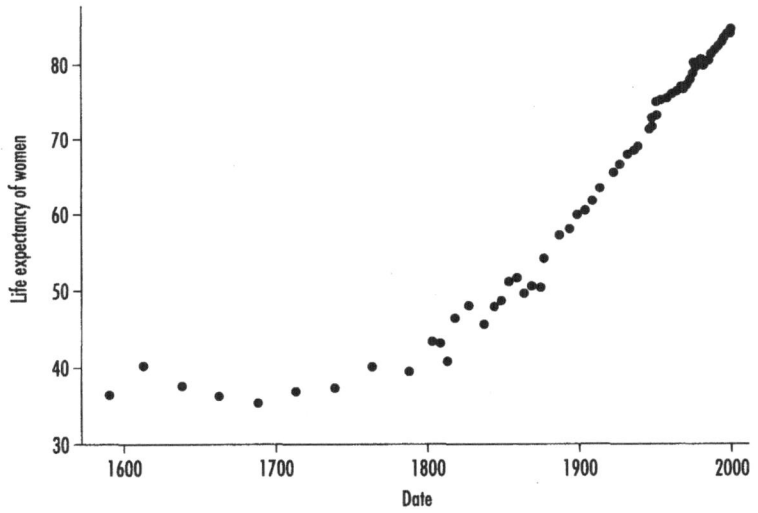

Figure 2.1 Life expectancy of women in country of highest life expectancy. Adapted from Oeppen, J. and Vaupel, J. W. (2002) *Science*, **296**, 1029–31.

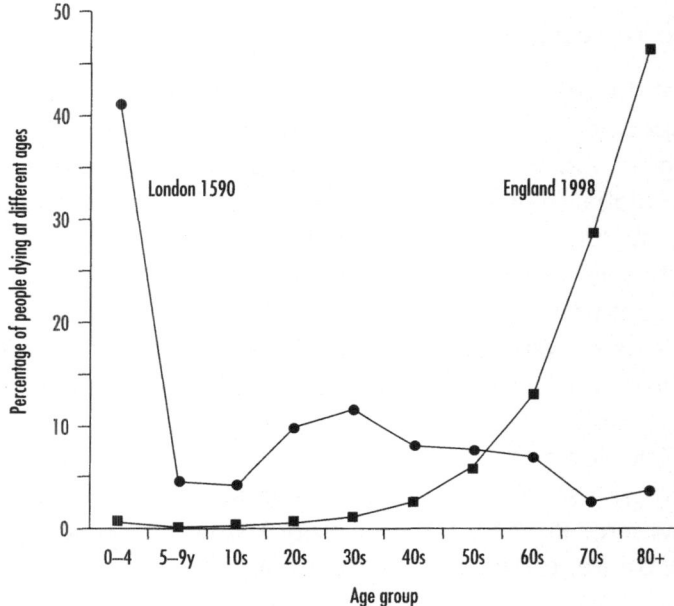

Figure 2.2 Percentage of total deaths in different age groups. Data for the London parish of St Botolph 1583–99, and for England and Wales 1998. Adapted from[7].

It is hard to underestimate the impact of this 'mortality transition' on the life experience of the average person (as well as billions of real people). Before 1800 most people died before reaching adulthood, with considerable psychological, social and economic costs. Life expectancy of pre-modern populations, from the Stone Age, through Classical civilization, to the eighteenth century, was around 30 years[5]. The largest toll of death was that exacted in infancy and childhood: about 20% of newborn children died in their first 12 months and another 20% before they reached five years of age (Figure 2.2). How long an individual could be expected to live was unpredictable, which must have been unsettling for the individual concerned, but also for those caring for or being cared for by that

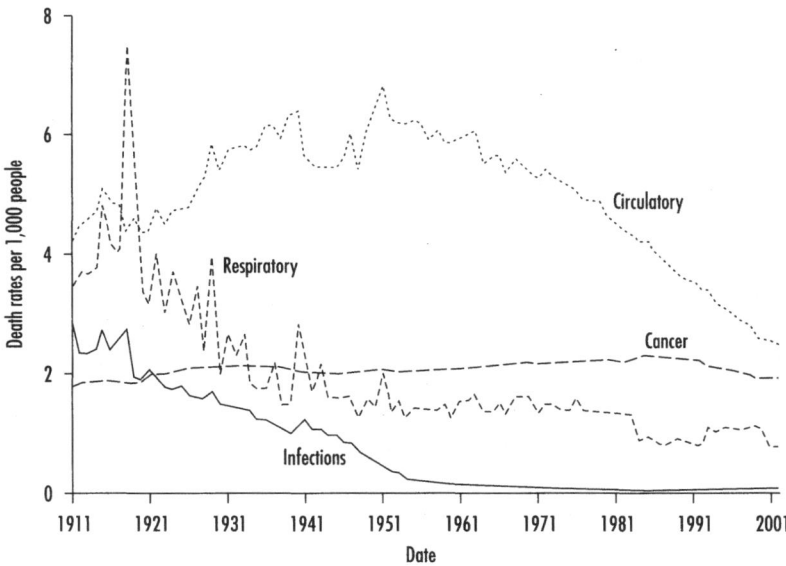

Figure 2.3 Age-standardized mortality rates for selected broad disease groups, England and Wales. Office for National Statistics, UK, 2004.

individual. And a lack of middle- and old-aged people meant an absence of experience that individuals and society could draw on; for instance it was rarely possible for adults to consult their parents.

Because so many people died in childhood prior to the nineteenth century, it has been thought by some that this accounted for most of the difference in life expectancy between now and then, so that once people escaped the dangers of childhood they had a reasonable chance of reaching a ripe old age. Figure 2.2 shows that in sixteenth century London, almost half of those born died in childhood, but of those who did survive the first ten years of life, subsequent life expectancy was still only about 40. The commonest age of death of those surviving childhood was in the 30s. Only about 5% of people died older than 70 years, whereas by 1998, about 75%

of people died older than 70 years. Reaching 70 years of age in the sixteenth century was rare but not exceptional. Note, however, that reaching 85 years of age was more-or-less unknown, whereas by 1998 it was common for females in England. In premodern cultures, adults who had survived the perils of childhood could still only expect to live to about 40 years old on average[1]. As the French humanist Michel de Montaigne wrote in the sixteenth century:

To die of old age is a rare, singular and extraordinary death, and so much less natural than others: It is the last and extremest form of dying.

In the US the percentage of the population over 65 was 4% in 1900, 8% in 1950, 12% in 2000, and projected to be 20% in 2050. More dramatically, the percentage of the US population over 85 was 0.1% in 1900, 0.4% in 1950, 1.5% in 2000, and projected to be 5% in 2050[2]. Thus there are 15 times as many very old people as there were in 1900 as a proportion of the US population, and by 2050 there are expected to be 50 times as many. Similar figures can be found throughout the developed world. Thus it is not true to believe that old people were just as common in the past – very old people are a new and 'unnatural' phenomenon.

This change during the nineteenth and twentieth centuries due to reduced death rates at all ages is known as mortality transition[3]. It might be thought that this transition was restricted to developed countries in the West. In fact, although the mortality transition occurred later in developing countries, it occurred or is occurring more rapidly there. So by the end of the twentieth century, life expectancy in most of the developing world was within 10 years of that in the developed world, and closing. On the other hand there are undoubtedly areas of the world where life expectancy is low, and getting lower.

But the big picture, over larger areas and times, is that life expectancy is rising almost everywhere. Even in Africa, the continent with the lowest figures, life expectancy jumped from 38 years in 1950 to 51 years in 2000.

The mortality transition has been attributed to improvements in six areas: public health (e.g. sanitation and vaccination), medicine (e.g. drugs such as antibiotics), economics (countries and people got richer), nutrition (reduced famine and malnutrition), behaviour (individual and family management of health), and education (literacy and schooling). The truth is probably a mixture of these six in different proportions at different times and in different regions of the world. The underlying causes of the transition are an increase of knowledge (both by the individual and society) and of wealth (which enabled that knowledge to be put into effect).

Life expectancy at birth is the number of years of life expected of a newborn baby on the basis of *current* mortality levels for persons of all ages. It is a summation of the current probabilities of survival at each year of life. It's actually rather a subtle measure based on the statistics of death at a particular place and time, and is difficult to relate to an individual's lifespan, but does give an indication of average lifespan. Importantly, however, because life expectancy is rapidly increasing, current life expectancy figures substantially underestimate how long the average person born today is likely to live.

Today life expectancy continues to rise at a rapid rate (Figures 2.1 and 2.4). In 2000, average life expectancy in the world was 67, but by 2050 the UN estimates it will be 76. That is, in 50 years the average lifespan is expected to increase by almost 10 years; this is the same rate of increase of lifespan that has been occurring over at least the last 100 years (Figure 2.1). The current record holder is Japan, where female life expectancy is 85 years, and on current trends will be 97 years by 2050.

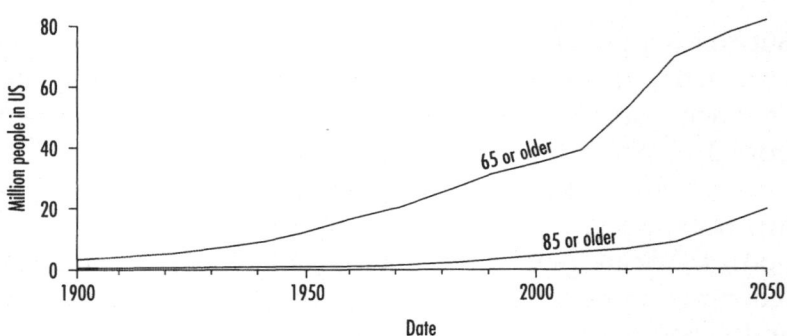

Figure 2.4 Total number of persons aged 65 and older or 85 and older in US in millions. Numbers after 2000 projected by US Census Bureau.

Immortality, however, is not the same thing as eternal youth. This was the fatal confusion that Tithonus made. Although life expectancy has risen and continues to rise rapidly, aging and the rate of aging have remained almost unchanged.

Flavours of death

What were the causes of death in the past, and what are they likely to be in the future? These questions are not as simple as they may seem. Indeed, the whole concept of 'cause of death' is doubtful, and has changed radically over history and differs between different cultures.

For many periods and religions the cause of death was God's will or Man's sin. With the rise of medicine and science the cause of death became a disease caused by a malfunction in the body (such as blockage of a blood vessel) or an external pathogen (such as a bacterium). But if the blocked blood vessel that caused death was in turn caused by degenerative disease, which in turn was caused by aging, is the real cause of death 'old age'? And if the cause of death is aging itself, is this functional or malfunctional? All causes must in turn be caused, so we shouldn't be looking for a single event but rather a chain of events. And that chain of events radiates out into the environment and back into history. And all causes act within a context that may be thought of as ancillary causes. The bacterium may have killed the man, but why did it kill this man and not others? Should we blame the bacterium, or his nutrition, the public health system, his doctor or the economic system? Was his immune system compromised by stress, pollution, poverty or genetics? Was the bacterium, stress, pollution, economics or genetics the true cause? The cause of death is never simple.

Dramatic changes in the causes of death and how death is attributed are evident in the list published by William Black in 1788 (Table 2.1, compared to a recent British list). Black compiled the causes of death of 1.7 million Londoners during the period 1701–76 using bills of mortality. The list consists mainly of what we would now regard as signs and symptoms of disease rather than the 'causes', i.e. the diseases themselves. The

Table 2.1 Leading causes of death in London, 1701–76, and England and Wales, 2003[4].

London 1701–76	% of deaths	England and Wales 2003	% of deaths
Convulsions	28	Cancer	26
Consumption	17	Heart disease	25
Fevers	15	Stroke	11
Smallpox	9	Respiratory disease	14
Age	8	Digestive system	5
Teeth	5	Accident and injury	3
Dropsy and tympany	4	Nervous system	3
Asthma and tissick	2	Endocrine system	2
Colic and gripes	2	Mental disorder	2
Childbed	1	Infectious and parasitic	1

symptoms (including convulsions, consumption, fever and smallpox) point to acute infectious diseases as being the main cause of death in eighteenth century London. The main cause of consumption (wasting) was tuberculosis (TB).

Today infectious diseases are a rare cause of death in the West, although contributing to respiratory disease. In the modern list, death is mainly attributed to diseases of the internal organs. The difference between the two lists partly reflects the fact that in the eighteenth century the cause of death was read from signs and symptoms on the *surface* of the body, whereas during the nineteenth century autopsies delving within the post-mortem body became more common, so that death became associated with changes in the internal organs, as exemplified by the modern list.

Black's list includes a variety of apparently rather exotic causes of death, such as convulsions (fits), consumption (wasting), dropsy (excessive swelling of tissues), tympany (swelling of the abdomen with gas), colic and gripes (abdominal pain), and childbed (following childbirth). These are rough translations, but if we go back to the seventeenth century it was possible to die of even stranger maladies for which there are no obvious, modern equivalents, such as Jaw-faln, Rising of the Lights, Livergrown, King's Evil, bloudy flux, impostume or being blasted.

The systematic classification of diseases (nosology) started with François Boissier de Sauvages's *Nosologie Methodique* (1772) in France, and in Britain William Cullen's *Nosology: or a systematic arrangement of diseases by classes, orders, genera and species* (1785) using Linnaeus's system of classification. In 1839 the Registrar General for England and Wales published their First Annual Report documenting for the first time the likelihood of death at different ages and the causes of death for a whole nation. The application of experimental science to investigating causes of disease resulted in new concepts, such as the 'germ theory' of infectious disease, mainly by Louis Pasteur in the 1860s. This theory itself contributed to the decline in infectious diseases, and replaced the older miasmic theory that attributed disease to odours from putrefying matter.

Science changed the representation of death in the eighteenth, nineteenth and twentieth centuries in many other ways. The gathering of death statistics, application of probability theory and publication of 'life tables' led to a kind of actuarist's vision of death: death that was abstract and predictable[5]. This depersonalized view of death is illustrated by Joseph Stalin's chilling comment: 'A single death is a tragedy, a million deaths is a statistic'. This contrasts with the 'untamed' death of the Middle Ages: death that was random, feckless and capricious.

The application of science to death in the modern age led to a process of rationalization: the expulsion of notions of mystery, wonder and visceral fear, and the reduction of events to their technical and controllable forms. Death was tamed by reducing it to a physical event with a single cause locatable within the body. The present-day cause-of-death certificate is based on five underlying assumptions about death: (1) it is a product of pathology; (2) it is a physical event; (3) the cause is a visible thing; (4) the cause is a single event; and (5) the cause is present at the moment of death. Several commentators[6] have pointed out these assumptions are, at best, unrealistic, but lead to the illusion of control. As sociologist Lindsay Prior[5] says:

> in the modern world, the explanation of death (one of the great imponderables of life) in terms of distinct and limited number of disease forms helps to generate the illusion that death can somehow be controlled.... Death it seems stalks only those who are careless – this despite the fact that the mortality rates of those who are careless of their

health and those who are fastidious remain stubbornly similar (viz. 100 per cent).

The convenient fiction of attributing death to a single cause in order to fill out the death certificate can lead to other fallacious conclusions from death statistics. Old people appear to die from vascular disease, cancer or pneumonia; yet the incidence of these diseases rises dramatically with age, suggesting that the diseases themselves are caused by aging. Many doctors would now accept that it is more honest to say that old people die from old age, rather than a particular disease, but 'old age' is not acceptable as a cause of death on the death certificate.

The whole enlightenment concept of 'disease' as the cause of pathology and death is now under threat. Diseases may be thought of as symptoms of more distal causes such as the individual's genetics or environment. For example, an individual's heart disease may be attributed to their 'susceptibility genes', their diet, their obesity, their smoking, their age or their culture. To refer to these more subtle causes, the concept of a 'risk factor for disease' was introduced – thus smoking is a risk factor for heart disease (rather than a cause of death).

Many conditions, such as Parkinson's disease, that were thought to be homogeneous entities are turning out to be several or many different diseases with similar symptoms. Some 'diseases', such as epilepsy or schizophrenia, may be symptoms alone, with no common underlying disease at all. For the last two hundred years doctors have sought to identify particular diseases within their patients and then treat those diseases. In the future, doctors will need to focus more on either the symptoms themselves or on more distal causes such as lifestyle and genetics. Many doctors and governments have already changed their focus in this direction, but there is still a long way to go.

Historians have tried to examine the causes of death during many different periods and regions of the past. There is a general consensus that the main causes of death over the whole of the historical period up until the mortality transition were endemic infectious diseases. These included: smallpox, gut and respiratory infections, influenza, malaria, measles, diphtheria, typhus, plague, tuberculosis and bronchitis, among others. Black's list includes many signs and symptoms of these diseases, such as fever and convulsions. Epidemics, famine and war were also important and could temporally and locally dominate death statistics. For example, the Black Death killed about one-third of all Europeans in just three years (1347–1350).

To die from an endemic infectious disease, epidemic, famine or war was not nice. But it was relatively quick compared to death by the chronic degenerative diseases of internal organs that now dominate. The main infectious diseases could not be effectively treated before the introduction of modern antibiotics in the 1940s, and the consequent death lasted days or weeks, rarely months. Such death may indeed have been more dramatic and traumatic than modern death, because it was rapid and happened to relatively young people. But modern death by cardiovascular disease, cancer or neurodegenerative disease lasts months or usually years, and treatment may prolong this. These are degenerative diseases that develop slowly and inexorably with age, and merge into the process of aging itself.

Increasing life expectancy has not just affected how people die, but also how they live, in particular the extent to which they live with chronic disease and disability. Older people are more prone to chronic diseases, such as arthritis and rheumatism, osteoporosis, high blood pressure and heart failure, cancers, diabetes, depression, varicose veins, ulcers, and strokes; they are also more susceptible to acute illnesses such as respi-

ratory infections, falls and other accidents; and they are also
much more likely to have disabilities such as deafness,
blindness and immobility.

In the UK in 1990, about 1% of 65–75 year olds were blind
and 1% deaf, but about 5% of those over 75 years were blind,
5% were deaf, and almost 20% were housebound[7]. A quarter
of those over 60 years have some degree of macular degeneration of the eyes. Osteoporosis (bone thinning and loss)
afflicts postmenopausal women, causing disability via frac-
tures of vertebrae, wrists and hips. Half of all American hospital beds for trauma patients are occupied by hip fracture
sufferers, and osteoporosis is the twelfth most common cause
of death in the US. Old people are also dramatically more
likely to develop neurodegenerative diseases, so that the current and predicted increase in the old and very old population
(Figure 2.4) is the cause of the current and predicted increase
in the population with Alzheimer's (Figure 2.5). A staggering
12 million people in the US alone are predicted to have
Alzheimer's in 2040.

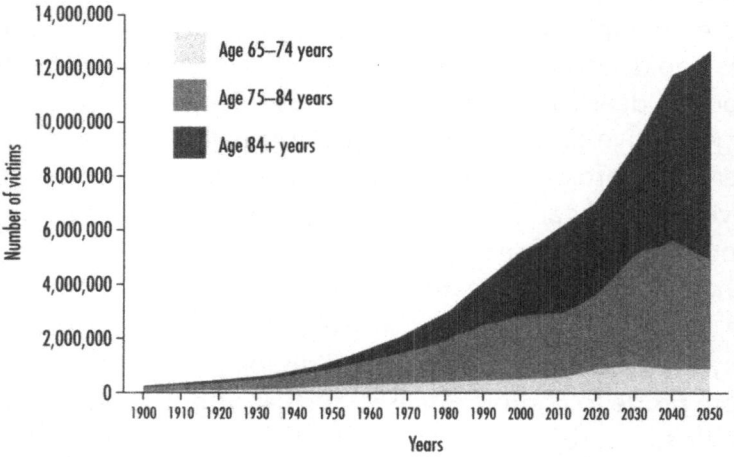

Figure 2.5 Prevalence of Alzheimer's disease at different ages in the US. Numbers after 2000 projected by
Hebert, L. E. *et al.* (2003) *Archives of Neurology*, **60**, 1119–22.

Before the modern era, most people died before they were old, and therefore chronic disease and disability were relatively rare and took milder forms. It is true that 200–2000 years ago many people lived with chronic endemic disease and disability. But in general people lived with these for relatively short times, while the degenerative diseases and disabilities of aging that afflict the modern age are by definition chronic and progressive. And chronic illness is still on the rise: 20% of people in the UK in 1972 reported having a long-standing illness, whereas by 1998 this figure had risen to 34%, i.e. one third of the population[1]. Part of this increase is due to medical advances keeping people with chronic diseases such as heart disease alive longer. Part is due to increased longevity, allowing people to live long enough to suffer from age-related disability and degenerative diseases. The result is the rise and rise of chronic disease.

We have looked at two contributions to the change from digital death in the past to analogue death in the present and future: (1) most people died from untreated infectious diseases in the past that killed rapidly, whereas today we die from degenerative diseases that develop over years and kill slowly; and (2) most people died young in the past before they had time to age, whereas today we die old or very old after we have aged for decades. In the next chapter we turn to how death is actually experienced now.

interlude 2
the meaning of death

What is death? We have already dealt with various definitions of death, but here I am concerned with the difference between the state of being dead and the state of being alive. So this is equivalent to asking: what is life?

Biologists in the eighteenth and nineteenth centuries got bogged down in this question, to the extent that they reified life as a force, which they (e.g. von Liebig) called the vital force. These 'Vitalists' were dualists in the sense that they believed that matter and the life force/vital spirit were two entirely different substances that could not be explained in terms of each other. This distinction between matter and spirit goes back at least as far as Galen, who distinguished between Natural Spirits (from the liver), Vital Spirits (from the heart) and Animal Spirits (from the brain), which 'animated' matter.

In opposition to these Vitalists have been various types of Materialists, who have maintained that spirit and essence do not exist; all can be explained in terms of the arrangements and processes of matter. The Materialistic tradition stretches from the Greeks Democritus and Epicurus, via Descartes and Lavoisier, to modern science. Ultimately it was found that the concept of the vital force was scientifically sterile, as it was shown that more and more supposedly vital processes such as movement, respiration and sensitivity could be modelled with non-living machines. And the components of our living bodies looked more-or-less like their non-living counterparts.

One celebrated milestone in the demise of vitalism was Luigi Galvani's ghoulish experiments in the 1770s showing that a dismembered frog's leg could be made to contract by electricity from a lightning conductor. Another was the chemical synthesis by Friedrich Wöhler in 1828 of the supposedly living substance urea from non-living materials. Thus it was slowly revealed that there was no inherent difference between the substance and processes of living matter and those of dead matter. Living matter was simply dead matter, differently organized (and vice versa).

But then we are left with the thorny question as to *how* life is differently organized compared with death. And to this question we still do not have a really satisfactory answer. The essential differences between living and dead matter are reproduction and self-organization. But the distinction is always being challenged by devising machines that can do these things. However, we no longer believe that it is theoretically impossible to build a living thing – it is one of the current aims of systems biology – but in practice we have no clear idea as to how to do it. Systems biology has arisen in the last 10 years as a reaction to reductionist biologists taking biology apart but never attempting to put it back together again. Everything in the body has been taken apart and analyzed *ad infinitum*, but no one has any idea how to put these components together again, to make a whole cell or live body. Some systems biologists are trying to redesign life and rebuild it from scratch. But it would be helpful to know what life is.

Death too has inevitably been reified and anthropomorphized as a thing or person: a malevolent spirit, a vengeful god, the four horsemen of the Apocalypse, or a hooded skeleton patiently awaiting us, scythe in hand. But death is not a thing – it is the absence of a thing: life. And life is not really a thing – it is set of processes and functions such as moving, breathing and thinking. So death is the absence of these functions, and dying is the transition between being fully functional and the complete absence of these functions.

One place to test our concepts of life and death is at the margins or interfaces between the two. The origin of life is one such interface where dead things became alive by self-organization and reproduction. Eventually the information required to reproduce this trick became stamped in one common medium: DNA. Nowadays we recognize as alive anything that is a descendant of those historical events, i.e. anything that now carries DNA and derived that DNA via the generations from the origin of Life. However, there are entities that lack DNA, such as prions, RNA viruses or red blood cells, which are difficult to classify as alive or dead, but are perhaps best thought of as components or derivative of life. And there are constructs that lack anything biological, like the Mars Rovers or the Internet, but which we might at least consider as derivative of life.

Consciousness is a central concern of us humans in distinguishing ourselves as alive from corpses or fancy robots/computers that are dead. Thus we get a modern definition of death, in the sense of dying, as the permanent loss of consciousness. According to this definition there is no point worrying about death, because when dead we won't be aware of it.

3
kicking the bucket

Death and plumbing

Almost half of all people dying today die from some problem with their plumbing. The blood vessels of our vascular system fur up, lose their elasticity and break down, blocking the essential blood supply to the heart and brain. These are some of the most common forms of death in both the developed and developing worlds, far deadlier than the more newsworthy AIDS or Alzheimer's. You will probably die from a vascular disease.

According to legend, the Greek hero Achilles, the bravest, handsomest and greatest warrior of the Trojan wars, was dipped as a baby by his mother into the River Styx in order to make him invulnerable to death. Unfortunately she held him by his heel, which left him vulnerable to an arrow fired by the Trojan prince Paris, and guided by Apollo to its deadly target. The heel and Achilles tendon was one part of the body generally left unprotected by armour in Ancient Greece. Times have changed, and our Achilles heel now lies deep within the body, in the walls of our blood vessels.

The root cause of virtually all vascular disease is the hardening, narrowing and furring up of the arteries with age – a process called arteriosclerosis. 'Arterio' refers to the arteries – the large blood vessels that generally carry the blood from the heart to the various organs of the body; and sclerosis simply means hardening. The cause of this hardening and narrowing is the formation of fatty deposits on the inside of the vessel wall, called 'plaques' or 'atheroma'. Hence the alternative name 'atherosclerosis' for the hardening of blood vessels due to the formation of these fatty deposits.

Edward Jenner, who introduced the smallpox vaccination in 1798, also observed the effects of arteriosclerosis in his patients. Jenner autopsied as many of his deceased patients as he could. During one autopsy he noted the effects of arteriosclerosis on the coronary arteries of the heart:

My knife struck something so hard and gritty as to notch it. I well remember looking at the ceiling, which was old and crumbling, conceiving that some plaster had fallen down. But on further scrutiny the real cause appeared: the coronaries were become bony canals.

Arteriosclerosis is not just common, it is universal. A famous study on young soldiers killed during the Korean War found signs of arteriosclerosis in almost all of them. It is a gradual process, starting in adolescence and progressing with age to different degrees in virtually all adults. The extent and rate of its progression, is determined by diet, smoking, exercise and genetics. And the many routes from sclerotic arteries to death are long and tortuous, involving a plethora of unpleasant symptoms and pathologies. While arteriosclerosis itself rarely appears on the death certificate, it leads to death by two main routes: heart disease due to blockage of the coronary arteries, or stroke due to blockage of arteries supplying the brain.

Coronary heart disease, also known as ischaemic heart disease, is the biggest killer in the western world. The coronary arteries are the blood vessels supplying blood to the heart. Constriction or blockage of the arteries supplying the heart due to arteriosclerosis results in 'ischaemia' i.e. insufficient blood supply. The heart needs a continuous supply of blood to replenish the oxygen that it is consuming at a rapid rate. The oxygen is used to burn food, releasing energy that drives the heart beat and the pumping of blood around the body. The faster the heart beats and the harder it works the more oxygen has to be consumed by the heart. During exercise the oxygen consumption of the heart increases up to fivefold, while smaller increases occur during emotional stress.

The more the coronaries are narrowed and blocked by fatty plaques the less capable the heart is of working harder. This mismatch between oxygen supply by the coronaries and oxygen demand by the heart eventually leads to angina, a searing chest pain. This may occur regularly for many years without further symptoms, but it may also announce a heart attack. Half of all coronary heart disease ends in sudden death by heart attack; the rest progresses to chronic congestive heart failure.

It has been said that death by heart attack approximates the ideal way to go. The victim may have had no previous symptoms up to the moment of the heart attack, and will be dead within minutes. This is a bolt-from-the-blue death, typically hitting the older person when exercising or stressed. However, only about one in five people dying from heart disease die in such (relatively) ideal circumstances – that is from their first heart attack having had no previous symptoms. Most have had years of previous symptoms such as angina, and many have had a series of previous heart attacks. Each attack kills a little bit of the heart muscle, making it weaker and more prone to another attack.

Death by heart attack is becoming less common in the developed world. There has been a 30% drop in the past three decades (Figure 2.3) – due to a reduction in risk factors such as smoking, high blood pressure and cholesterol as well as improved medical treatments. However, partly as a consequence of preventing acute heart deaths, death by chronic congestive heart failure has increased.

Heart failure is a result of the increasingly scarred, weakened and sclerotic heart's inability to pump out sufficient blood with each beat from the veins to the arteries. Lack of arterial blood causes muscle fatigue and other organs lose function. Insufficient clearance of venous blood causes a build-up of blood pressure in the veins and organs, pushing blood plasma (mainly water) out into the tissues. Increased water results in swelling of the tissues, further increases pressure and congestion, and decreases oxygen supply to tissues. Organs such as

the kidneys and liver are unable to function properly due to
lack of oxygen or excessive swelling. The lungs may literally
drown in excess fluid, making breathing difficult. The blood
circulation increasingly sufferers from congestion, like a city
strangled by traffic. The heart compensates by getting larger
and thicker, but this itself causes more problems because it
requires a larger blood supply to feed the larger heart. However, this larger blood supply is not available due to sclerotic
coronaries, and the result is more failure. The victim of chronic
heart failure eventually spirals out of the doctor's control, and
may die from failure of the lungs, kidney or liver rather than
the heart itself.

Prevalence of heart failure increases steeply with age.
Around 1% of men and women aged under 65 have heart failure, increasing to about 7% of those aged 75–84 years and
15% of those aged 85 and above[1]. Heart failure is a major
cause of death. But again because of medical advances it is
being converted from an acute to a chronic form, condemning victims to live for years with symptoms such as extreme
breathlessness, fatigue and exercise intolerance. Lynne
Stevenson, an expert on heart failure from Boston, recently
summarized the situation[2]:

> Twenty years ago, patients with heart failure usually died
> suddenly as the disease progressed, such that there were
> few patients suffering with refractory symptoms. As dis-
> cussed above, fewer patients now die suddenly with heart
> failure [so] increasing efforts are required to help them and
> their families face the end with comfort.... Heart failure is
> now truly a chronic disease.

So the good old-fashioned 'heart attack' turned into 'heart
failure', and is slowly turning into a long, drawn out 'heart
retreat', ending no doubt in 'heart defeat'. Increased survival

with heart disease is a good thing, but it would be a much better thing if heart disease were cured rather than converted into a chronic condition.

Stroke and vascular dementia

> Her unseeing eyes had the dullness of oblivion; her face was expressionless. Even the most impassive of faces betrays something, but I knew at that instant of absolute blankness that I had lost my grandmother. I shouted, 'Bubbeh, Bubbeh!', but it made no difference. She was beyond hearing me. The cloth slipped from her hand and she crumpled soundlessly to the floor.

In this quote from Sherwin Nuland's excellent book *How We Die*, Nuland, a surgeon, describes the impact of a stroke on his beloved grandmother Bubbeh. His subsequent account of her death in harrowing prose conveys something of the miscellaneous incomprehensible indignity of a modern death. Stroke is overwhelmingly a disease of old age, and a common cause of death in the aged.

Stroke (apoplexy) is the brain's equivalent of a heart attack. It is caused by blockage of one or more arteries supplying blood to the brain, resulting in death of the most important brain cells: neurons. Stroke is the third most common cause of death in the developed world, instigating one in ten of all death. Approximately half a million US citizens per year have a stroke. Of these one third die, one third are left with a permanent severe disability, and one third will recover but are at risk of a subsequent stroke.

There are three different causes of a stroke. The build up of fatty plaques on the inside of brain arteries may trigger a blood clot ('thrombus') to form on the vessel wall that blocks the vessel. Alternatively, part of a plaque may break off the vessel wall, forming a clot within the bloodstream, known as an 'embolus' (Greek for plug). This may then literally plug an artery in the brain, preventing further flow. Thirdly, and most dangerously, high blood pressure (hypertension, which is again caused by arteriosclerosis) may rupture the weakened vessel wall, causing blood to burst through (a haemorrhage).

Whatever the cause of a stroke, the result is insufficient blood supply to the brain, and rapid death of neurons. The symptoms that follow depend on just which artery is blocked and thus which part of the brain is knocked out. Most commonly it is one of the middle cerebral arteries. These supply areas of the cerebral cortex controlling hand and eye movements, hearing and language, as well as more general functions such as perception, thought and volition. A stroke that does not kill can in some cases be cruel, permanently wiping part of the self.

Larger strokes result in coma, which may be accompanied by complications such as reduced blood pressure or heart function, exacerbating the damage. Coma can bring on respiratory infections in elderly patients by eliminating defence mechanisms such as the coughing reflex. That is why death by pneumonia is common in comatose stroke patients. If the stroke damage is great enough the brain swells and pushes down through the base of the skull, physically damaging the brain stem. If this happens death usually ensues rapidly, as the brain stem controls vital functions such as the heartbeat and breathing. But the ways in which stroke can kill are as diverse as the body functions that the brain controls, including induction of diabetes, increased blood pressure, or paralysis of the chest wall preventing breathing.

A stroke is an acute, dramatic event equivalent to a heart attack, but the elderly may also suffer from multiple, small strokes or 'transient ischaemic attacks'. These result in little or no immediately obvious symptoms, but over time damage to the brain accumulates, eroding the intellect. Walter Alvarez, a renowned Chicago clinician quoted 'a wise old lady' who said to him: 'Death keeps taking little bits of me'. And he noted that:

> She saw that with each attack of dizziness or fainting or confusion she became a little older, a little weaker, and a little more tired; her step became more hesitant, her memory less trustworthy, her handwriting less legible, and her interest in life less keen. She knew that for ten years or more, she had been moving step by step towards the grave.

In 1946 Alvarez drew attention to mini-strokes as a cause of dementia, after watching the insidious process in his own father. It was first noted by Alois Alzheimer in 1899, several years before he described the different form of dementia that bears his name. Roughly 10% of senile dementia cases may be caused by this type of 'vascular' dementia, which is yet another manifestation of arteriosclerosis. The interface between vascular dementia and Alzheimer's disease is now an active area of medical research. One form of dementia may cause the other, and there are intermediate forms.

Deaths caused by stroke (per thousand people at any particular age) have declined steadily since the beginning of the twentieth century. Since 1950, these deaths have fallen by 70% in the US. Deaths due to cardiovascular disease (which includes heart disease and stroke) have declined 60% since 1950 and account for most of the decline in all causes of deaths during the same period in the US[3]. And cardiovascular death continues to fall rapidly – there was a 23% drop in the UK between 1997 to 2002 (Figure 2.3). Again, this welcome decline has the unfortunate consequence of replacing rapid forms of death (heart attack and stroke) with slower, more chronic ends (heart failure, dementia, cancer).

The cause of vascular disease in both the brain and heart is arteriosclerosis: hardening of the arteries. The cause of arterio-sclerosis, however, is unknown. That's quite a problem for something that causes nearly half of all deaths in the developed world. We know that smoking, obesity and the wrong kind of fat can all increase arteriosclerosis. Why is less clear. The current hypothesis is damage to the cells that line the blood vessels. Damage to these 'endothelial' cells may come about via turbulence of the blood, toxins, inflammation or 'age', but no one really knows. Once damaged, the endothe-lial cells allow white blood cells into the vessel wall, where they pick up vast quantities of fat and cholesterol, forming the fatty plaques. The vessel wall responds by becoming thicker and less flexible; blood flow falls and blood pressure rises. Now we are in trouble!

Inflammation in the vessel wall results in the production of oxidants, like hydrogen peroxide, that oxidize or damage the fat and cholesterol, which are then much more toxic to cells. Eventually the wall becomes overloaded with oxidized fat and starts to die. Collapse of the plaque may then cause a blood clot in the adjoining vessel, or part of the collapsed plaque may be carried away by the blood, forming a travelling clot that may block a vessel elsewhere. If the clot blocks a large vessel supplying the heart or brain, the result is a heart attack or stroke.

Respiratory disease and infections

Respiratory disease is a mixed bag of disorders that affect the ability of the lungs to get oxygen into the body. Respiratory diseases account for 10–15% of all death in the developed

world. The two main culprits are chronic obstructive pulmo-
nary disease (COPD) and pneumonia.

As its catchy name implies, COPD involves obstructed air-
flow in the lungs. COPD is an umbrella term that includes diseases such as bronchitis and emphysema, but these latter
terms are dying out. Whatever their problem, COPD patients
are continually short of breath, usually have a chronic cough,
and may cough up thick mucus. The disease results in progressive degeneration of the lungs, usually starting in the 40s, and
is largely irreversible and incurable. About 80% of cases are
caused by smoking, and the rest by pollutants, such as smog
or coal dust. These irritate the lungs, leading to chronic
inflammation that eventually destroys the lung tissue. The disease is worsened by lung infections, so antibiotics can help in
early disease, but there is no really effective treatment or cure.
COPD is on the rise as a cause of death and chronic disease in
the developing world, largely as a consequence of increased
smoking in developing countries such as China.

Before 1900 infections were the main cause of death. These
days in the developed world they apparently account for only
1% of deaths (Table 2.1 and Figure 2.3). However, bacteria
and viruses, in particular those causing pneumonia and influenza, contribute to respiratory disease. Respiratory diseases
are primarily caused by damage to the lungs, which are then
less able to resist infections. Micro-organisms in food, water
and surfaces can and have been effectively controlled by
hygiene and sanitation, but micro-organisms in the air can
travel long distances, from one lung to another, and are much
more difficult to control. Antibiotics have greatly reduced the
impact of bacterial respiratory infections, but susceptibility to
such infections increases with age. The reason that the old fall
victim to these bugs is that the immune system become less
and less effective with age at fighting infection. The result is
often pneumonia.

Pneumonia is a condition where the lungs' air sacs become inflamed, and collapse and flood with fluid and pus. Victims essentially drown on dry land. Pneumonia can result from an infection or physical or chemical damage to the lungs. It mainly kills children or old people. After her stroke (at the beginning of the last section) Sherwin Nuland's grandmother caught pneumonia. In fact, we now know that the immune system is severely depressed after a stroke, and many people die from the subsequent infections rather than the stroke itself. William Osler was of two minds about pneumonia in the elderly. In the first edition of *The Principles and Practice of Medicine*, he called it 'the special enemy of old age', but later he stated something quite different: 'Pneumonia may well be called the friend of the aged. Taken off by it in an acute, short, not often painful illness, the old escape those "cold gradations of decay" that make the last stage of all so distressing'. Unfortunately pneumonia too is on the decline, depriving the aged of their 'old friend'.

Sir William Osler, born in 1849, is one of the great icons of modern medicine. A professor of medicine first in Montreal, then Philadelphia, Baltimore and Oxford, he established the modern methods of teaching medical students (on the wards), and wrote their 'bible' (*The Principles and Practice of Medicine*). Towards the end of his illustrious career Osler gave a speech in which he approvingly quoted Anthony Trollope's novel *The Fixed Period*. The book envisaged a College where men retired at 67 and after a contemplative period of a year were euthanased. (Trollope wrote the book in his 67th year, the year of his death). In his speech Osler recommended that men be retired at 60 and 'peacefully extinguished' by chloroform at 61. He claimed that, 'the effective, moving, vitalizing work of the world is done between the ages of twenty-five and forty' and it was downhill from then on. This speech became infamous amongst gerontologists and the press, but may not have been entirely serious. Osler himself died at 70 years, during the influenza epidemic of 1919.

Such killers are making a comeback. The glorious victories of antibiotics and vaccines have made us complacent about infections. Antibiotic-resistant bacteria are spreading in our hospitals and contributing to the rise and rise of chronic disease. Tuberculosis (TB), caused by a bacterium infecting the lungs, was one of the main causes of death in the eighteenth and nineteenth centuries, and is returning now due to antibiotic resistance and AIDS. AIDS, spread by the HIV virus, is ravaging the developing world, while in the developed world it has been converted from an acute disease to a chronic disease thanks to drugs that contain but do not destroy the virus.

Cancer

Cancer is a relatively new phenomenon. Evidence of cancer has been found in Egyptian mummies from 5,000 years ago, but cancer was rarely recorded before the nineteenth century – for example, it is not evident in Black's list of the causes of death in eighteenth century London (Table 2.1). It reached public consciousness in the 1960s as the taboo plague, subsequently replaced in notoriety by Alzheimer's and AIDS in the 1980s. However, our persistent failure to conquer cancer, combined with the decline of other forms of death, means that it is now one of the commonest causes of death (Figure 2.3 and Table 2.1), with a lifetime risk of 1 in 3 of women and 1 in 2 of men in the developed world. The causes of cancer are still unclear, but age is the most potent of all carcinogens. The incidence of cancer rises exponentially with age. This is why cancer was rare in the past. And this is why, as the population ages, and other forms of death decline, cancer is set to become the most common cause of death (Figure 2.3).

Cancer results from changes in the genes controlling cell division, cell migration and cell death. Causes of these mutations include noxious chemicals, radiation, viruses or just random errors when the cell is copying its own DNA. Several mutations in different genes must accumulate over years or decades before ordinary, law-abiding cells become killers.

The human genome is 3 billion letters long: the chances of accumulating a particular combination of changes by random mutations is astronomically small. However, there are also an astronomically large number of cells in the human body, perhaps 100 trillion (actually, we only have a rough idea). And it takes only one to produce a lethal cancer. This shortens the odds somewhat.

It still takes a long time to accumulate the relevant mutations: that's why your chances of developing cancer increase with age. On the other hand the cancer mutations themselves can speed the accumulation of further mutations. For example they can increase the production of toxins that cause mutations, or stimulate cell division during which copying errors occur, or eliminate processes that correct DNA errors. So the accumulation of errors can snowball out of control.

The vast majority of errors either have no effect or lead to cell death. Even if they survive their own error catastrophe the tumour cells are likely to be attacked and removed by the body's immune system. Only a very particular set of errors leads to a cell dividing rapidly (despite its neighbours telling it to stop), procuring a new blood supply, and invading other tissues via that blood supply, but avoiding self-destruction and immune attack. Amazingly all of this is achieved by a process of random mutations: the cancer cell, or ordinary cell from which it evolved, never set out to cause damage to the body – it just arose from a catalogue of mishaps.

The development of cancer cells from ordinary cells is speeded up by the process of evolution by natural selection. A cancer cell evolves from an ordinary body cell during the lifetime of a person by the gradual accumulation of mutations. At each stage in their evolution the majority of cancer cells may be eliminated, either by the body or failure of the cancer cell to survive. Each new mutation that promotes survival, reproduction, defence or spread, will also promote the survival and spread of the mutated genes within the dividing cancer cells that carry those mutations.

To thrive, a cancer needs to accumulate mutations (that are helpful to the cancer) in roughly half a dozen genes out of the 30,000 in the human genome. Different cancers have different sets of mutated genes, because there are hundreds of genes that affect the ability of a cancer to thrive. Nevertheless there are mutations in particular genes that are particularly helpful to a developing cancer, and therefore are commonly found in cancer patients' tumours.

For example, the gene called p53 is mutated in about half of all human cancers. This gene is known as the 'guardian of the genome', because if damaged DNA is detected then p53 orders the cell to stop dividing or commit suicide. p53 therefore prevents cancers from developing, and that is probably its main function. A cancer needs to disable this gene if it is to thrive. Another strategy is to hit genes that control cell division or cell suicide, so that the instructions from p53 cannot be executed. Many different genes that control division can be mutated in cancer; the mutations must be subtle because an increase in cell division is required rather than a simple loss of function.

Viruses can sometimes help. One nasty virus makes a mutated version of the EGF Receptor. Epidermal Growth Factor (EGF) is a local hormone that tells epidermal cells to grow and thrive by sticking to receptors on the cells. These activated receptors then send a message within the cell telling the DNA to use genes involved in growth and survival. In case you are wondering, an epidermal cell is a type of cell found of the surface of the body and internal organs. A normal epidermal cell is entirely dependent on its small local supply of EGF, so its growth is limited. If it leaves the spot where it is meant to be it loses its local supplier of EGF, and therefore stops growing and dies.

An epidermal cell infected with the nasty virus makes a mutated form of the EGF receptor. This defective receptor is fully activated all the time, even in the absence of EGF. So this cell gets the corrupted message to keep growing and dividing, no matter how many other epidermal cells there are or where it is. That cell is on its way to becoming a cancer. Incidentally, this helps the virus too by aiding the survival and spread of cells containing the virus. So viruses may carry a variety of other cancer-promoting genes that they originally acquired from our cells and then mutated to serve their own perverse purposes.

A virus is not usually sufficient to cause a full cancer. That requires a cocktail of different changes to the cell. One or two of those may be caused by a virus, others by random mutations, others by inherited mutations, and yet others by environmental toxins such as cigarette smoke.

All this gives the impression that cancer is one unified disease. In fact, it is a family of many different conditions, each with different causes, courses and consequences. The frustrating result is that medical lessons learnt with one cancer are rarely applicable to another. The most common cancers in the developed world are lung, gut, breast and prostate. Together these account for over half of all cancers. Even within one type of cancer, characteristics may vary wildly between different individuals, because the genetic mutations within individual cancers may vary and because the genetics of the individual patients are different.

Cancer deaths were remarkably stable during the twentieth century (Figure 2.3). A mild upward trend due to an aging population plus changes in diet and smoking was balanced by a mild downward trend due to advances in prevention and treatment plus changes in diet and smoking. Some cancers have gone up (e.g. kidney) while others have gone down (lung in men, breast in women)[4,5].

Cancer mortality is likely to decline very slowly over the next few decades, while cancer incidence may be stable or increase. In part this is due to more people surviving for longer with cancer. Survival with cancer, the time from diagnosis to death, has increased significantly in three of the most common cancers: breast, gut and prostate. For example, in the UK 5-year survival rates increased between 1970 and 2000 from roughly 20% to 45% for colorectal cancer, from 50% to 80% for breast cancer, from 30% to 70% for prostate cancer, and from 10% to 25% for myeloma[6]. This can be good news for the patient in terms of living longer, but it can also be bad news in terms of living longer with cancer; dragging death out over a longer period. In some cases those extra years granted are 'good' years well worth living. In many other cases those extra years are lived in the shadow of disease with multiple, invasive medical interventions, anxiety and disability. Cancer mortality is falling much more slowly than deaths due to heart disease and stroke, so that in the near future cancer will be the commonest cause of death (Figure 2.3). Unfortunately, as we shall see in the next chapter, cancer is not a nice way to die.

The rise and rise of chronic disease

The four main causes of death in the Western world are heart disease, stroke, infections and cancer. The age-adjusted incidence of each is stable or declining, but there is no prospect of a 'cure', perhaps because the 'cause' is the aging process. We are living longer with these diseases: they are therefore merging into the process of aging itself.

We have seen how each of these fatal diseases is being converted from an acute cause of death to a chronic condition:
heart attack into heart failure, stroke into vascular dementia,
epidemic to chronic infection, cancer death to cancer disability. In part, these disease conversions reflect the economics of
drug development by the huge pharmaceutical companies
that dominate Western medicine. It is much more profitable
to sell a drug that treats a disease rather than curing it – just
think of Zantac and Viagra. If a disease is cured the market dis-
appears and the millions invested in developing the wonder
drug are lost. A drug or treatment that completely cured AIDS
in one dose would be a financial non-starter for a drug company because of the huge costs of development and clinical
trials, whereas drugs that prevent AIDS from killing and convert the patient into having a lifelong dependency on those
drugs makes the companies involved some of the richest in
the world.

The conversion of acute to chronic disease, and consequent aging of the population, has also led to the rise and
rise of chronic disease. In the past there was relatively little
chronic disease and disability; now the average person experiences 10 years of it. And this figure is rising (see Figure 4.4).
The developed world is filling up with chronically ill old
people, who are relatively rich and thus form a lucrative
market (and lobby) for drugs and treatments that maintain
them in that state.

The rise of chronic disease is also a consequence of which
priorities have been identified and funded in medicine, medical research and public health. Patients and medics have, perhaps understandably, been more concerned to prevent death
than to prevent disease. Funding has followed those immediate concerns rather than taking a more long-term view. People
are more willing to give money to prevent death than to prevent disease. Young medics are more attracted to accident

and emergency medicine than to gerontology. Governments prefer to target death statistics than quality-of-life statistics. The result is the Tithonus scenario.

interlude 3
the search for immortality

One of the earliest accounts of death is that of the hero Gilgamesh, Lord of Uruk. Probably composed at the Sumerian court of King Shulgi in the twenty-first century BC, and recorded in Babylonian tablets from the seventeenth century BC. These words are old: far older than anything in Homer or the Bible. They come from close to the dawn of writing and civilization.

> The great wild bull is lying down, never to rise again,
> the lord Gilgamesh is lying down, never to rise again, ...
> he is lying on his death bed, never to rise again,
> he is lying on a bed of woe, never to rise again.
> He is not capable of standing, he is not capable of sitting,
>   he can only groan,
> he is not capable of eating, he is not capable of drinking,
>   he can only groan,
> the lock of Namtar holds him fast, he is not capable of
>   rising....
> For six days he lay sick,
> The sweat rolled from his body like melting fat....
> Great Mountain Enlil, the father of the gods,
> Conversed in the dream with the lord Gilgamesh:
> 'O Gilgamesh, I made your destiny a destiny of kingship,
>   but I did not make it a destiny of eternal life.
> For mankind, whatever life it has, be not sick at heart,
> be not in despair, be not heart-stricken!
> The bane of mankind is thus come, I have hold you,
> what was fixed when your navel-cord was cut is thus
>   come, I have told you.
> The darkest day of mortal man has caught up with you,
> the solitary place of mortal man has caught up with you,
> the flood-wave that can not be breasted has caught up
>   with you,
> the battle that cannot be fled has caught up with you,

the combat that can not be matched has caught up with
  you,
the fight that shows no pity has caught up with you.
But do not go down to the Great City with heart knotted
  in anger,
let it be undone before Utu,
let it be unravelled like palm-fibre and peeled like an
  onion!'[1]

This account of an individual's death is both strange and
strangely familiar. If we ignore the more mythological aspects
of this account, it tells of what seems like a familiar form of
death: a young man struck down in his prime and consumed
by a fever in a few days. Indeed it could be an account of
death caused by almost any infectious disease, probably the
commonest form of death at that time.

Gilgamesh is a semi-mythical hero, based on an apparently
real King Gilgamesh of the Sumerian city state of Uruk, living
(if he existed at all) around 2,800 BC. His story, which is one of
the oldest in the world and one of the most famous in antiquity, is in a sense itself the beginning of history, culture and
writing. Paradoxically, the Epic of Gilgamesh is still being constructed: thousands of fragments of it are scattered among
the world's storehouses, and thousands more lie below the
deserts of Iraq. The story had been laboriously copied countless times onto clay tablets by apprentice scribes of the ancient
cities of Mesopotamia, the Levant and Anatolia. 'The story'
exists in several different versions and in several different lan-
guages, from several different historical periods of two millennia of history. The main versions are Sumerian, Babylonian
and Assyrian, but none of the versions that we now have is
complete. What we have instead is a vast jigsaw puzzle of
eroded tablet fragments, most of which are missing, derived
from hundreds of copies of several different versions. Modern-

day scholars of ancient Mesopotamia are toiling away in the back rooms of museums and universities all over the world to translate those fragments, and every few years they publish new versions. The oldest story in the world is continuously born anew.

Appropriately, that story is about death and immortality. Gilgamesh begins the Assyrian version of the epic as the young, virile, war-mongering King of Uruk, lording it over his people:

> Gilgamesh sounds the alarm bell for his own amusement, his arrogance has no bounds by day or night. No son is left with his father, for Gilgamesh takes them all, even the children; yet the king should be shepherd to his people. His lust leaves no virgin to her lover, neither the warrior's daughter nor the wife of the noble; yet this is the shepherd of the city, wise, comely, and resolute[2].

But the untimely death of his passionate 'wild-man' friend Enkidu leaves Gilgamesh in a mid-life crisis. He is confronted by the prospect of his own mortality, which, however, he will not accept lying down. He claims to be 'two-thirds' a god, as his mother was an immortal goddess. However, his father, though a king, was mortal, and consequently his resentful son is destined to dust and oblivion. He abandons everything, and sets out alone for the ends of the earth in a doomed quest for immortality.

Gilgamesh follows the nightly path of the sun under the earth to the gardens of Paradise, and from thence he is ferried across the waters of death to the end of the world. There, at the end of his quest, he meets Utnapishtim, the one-and-only mortal that the gods have granted immortality. Gilgamesh asks him how he also may obtain everlasting life. But in reply Utnapishtim explains the impossibility of permanence:

There is no permanence. Do we build a house to stand for
ever, do we seal a contract to hold for all time? Do broth-
ers divide an inheritance to keep for ever, does the flood-
time of rivers endure? It is only the nymph of the dragon-
fly who sheds her larva and sees the sun in its glory. From
the days of old there is no permanence. The sleeping and
the dead, how alike they are, they are like a painted death.
What is there between the master and the servant when
both have fulfilled their doom? When the Anunnaki, the
judges, and Mammetun the mother of destinies, come
together, they decree the fates of men. Life and death they
allot but the day of death they do not disclose[2].

Utnapishtim then reveals a secret: long ago the gods became enraged by the clamour of humanity, and in a fit of
pique sent a genocidal flood to wipe the world clean. But Ea
(god of wisdom) warned Utnapishtim in a dream to save himself by building a great ark into which he should gather the
seed of all living creatures. Thus the world was reborn after the
flood, and the remorseful gods granted Utnapishtim everlasting life.
  Despite Utnapishtim's words of wisdom, Gilgamesh insists
that he be given the secret of immortality. Utnapishtim sets
Gilgamesh an instructive test: if he can defeat sleep, which is
regarded as the little death, for seven days and nights, then he
has some prospect of defeating the great death. Gilgamesh
fails miserably, and is sent home. But as a parting gift
Utnapishtim reveals the secret of a plant that restores lost
youth. Gilgamesh is triumphant, and at last heads home, but
on the way he loses the secret of everlasting youth down a
well. Oops! Gilgamesh is distraught and has to face his fate, to
return home empty handed and await his death. The Assyrian
version of the epic ends with Gilgamesh at last reaching
home, and consoling himself by contemplating the walls of

Uruk, the walls that he rebuilt, the testament that he believes will survive after his death to preserve his name for posterity, his little piece of immortality. It is these walls, the walls of Uruk, that were uncovered by modern archaeologists to reveal the remains of the oldest known city in the world.

The Epic of Gilgamesh is important partly because it reflects ancient concepts of death and immortality, and partly because it shaped such concepts. Mesopotamian ideas in turn influenced Jewish and thence Christian and Islamic beliefs. According to those Mesopotamian concepts death is a catastrophe, but it is unavoidable, and we must accept it. The blow is not softened by a sweet afterlife. Death is the oblivion of never-ending sleep, only broken by the nightmares of a shadow-world, where people's attenuated spirits survive as birds eating dust and clay in eternal darkness.

There are only two practical ways to obtain some degree of immortality: firstly through having children, and secondly through deeds that will live on in the collective memory of people. There was a particular concern that the individual's name should survive, preserved by descendants, by society or by writing. This may relate to the ancient belief that the representation of something (whether in image or name) was magically connected to the thing represented. Thus survival of someone's image or name enabled some form of survival of the individual. The final stanza of the old Sumerian story concludes:

Men, as many are given names,
their (funerary) statues have been fashioned since days of
   old,
and stationed in chapels in the temples of the gods:
how their names are pronounced will never be forgotten!
The goddess Aruru, elder sister of Enlil,
for the sake of his name gave men offspring:
 their statues have been fashioned since days of old,

and their names still spoken in the land.
O Ereshkigal, mother of Ninazu, sweet is your praise![1]

Ereshkigal was mother of death and goddess of the underworld. Men, who were mortal, could not escape becoming shades in her realm, but they could survive in two other forms: through the collective memory of their names, and through their offspring.

Gilgamesh complains to his comrade Enkidu at some point in the story:

I have not established my name stamped on bricks as my destiny decreed; therefore I will go to the country where the cedar is felled. I will set up my name in the place where the names of the famous men are written, and where no man's name is written yet[2].

Gilgamesh proposes to establish his name by killing the monster Humbaba in the land of the cedars (Lebanon), but Enkidu is afraid and Gilgamesh admonishes him:

Where is the man who can clamber to heaven? Only the gods live forever with glorious Shamash [the sun god], but as for us men, our days are numbered, our occupations are a breath of wind. How is this already you are afraid! I will go first although I am your lord, and you may safely call out, 'Forward there is nothing to fear!' Then if I fall I leave behind me a name that endures; men will say of me, 'Gilgamesh has fallen in fight with ferocious Humbaba.' Long after the child has been born in my house, they will say it and remember[2].

So Gilgamesh is telling Enkidu that it is pointless to fear death because death will come whatever we do, but if we do great deeds our names will endure even after we and our children are dead.

In the older Sumerian version of the story, quoted at the beginning of this interlude, Gilgamesh, who sought immortality in life, is instead granted immortality in death. At death the gods make Gilgamesh a minor god of the underworld, and in fact there was a Sumerian cult that worshipped Gilgamesh as an underworld god in the third millennium BC. But ultimately Gilgamesh's name obtained immortality through the story, the Epic of Gilgamesh. That story was famous for two thousand years throughout Mesopotamia, the cradle of civilization. The story was then lost for another two thousand years, before fragments were dug up in the middle of the nineteenth century from the ruins of Nineveh, where they had lain for 25 centuries. The name of Gilgamesh lives again!

4
death is falling apart

Physical, social and psychological death

Death is not just happening later and taking longer, it's also fragmenting into multiple forms. The experience of dying today merges into the general experience of aging, because the vast majority of people dying today are old or very old, and die from chronic, degenerative diseases that kill them over years or decades. And the disability and distress caused by these aging-related diseases merges with the more general aging-induced loss of functions and social disengagement. Social surveys have been used to investigate the prevalence of illness and disability in the population at various ages. The UK government's General Household Survey has shown a steady increase in the proportion of people reporting longstanding illness since 1972 and this is largely due to increased numbers of old people, as unsurprisingly the proportion of older people reporting such illness is much higher: about 70% in those over 75 (Figure 4.1). As a result of this increased level of disability, 30% of people aged 85 or more surveyed in the 1996 GHS needed help at home in climbing the stairs; 24% needed help

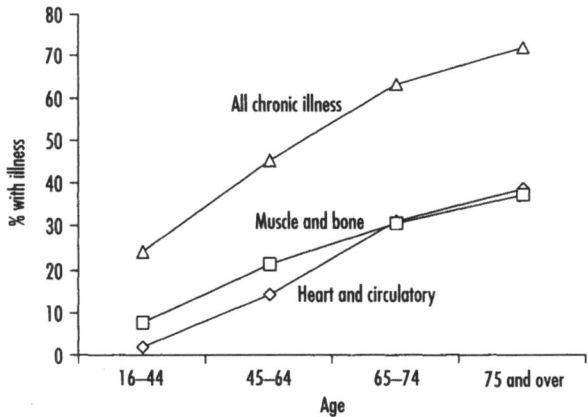

Figure 4.1 Percentage of the UK population reporting chronic illness of various types. Source: UK Office for National Statistics, 2004.

with bathing or showering; and 8% with dressing and undressing[1].

The same trends are evident in the US population[2], with 45% of 65–69-year-olds having a disability of some sort, 31% a severe disability and 8% needing daily assistance; increasing to 74% of those older than 80 having a disability, 58% a severe disability and 35% needing assistance. And these proportions are increasing, not because old people at a particular age are getting less healthy, but simply because people are dying older and older, so exposing them to illnesses that increase with age. The chronic illnesses reported by the aged include particularly: arthritis, hypertension, heart disease, cancer, diabetes and stroke, all of which increase with age[2,3].

So old age is on average characterized by chronic ill health and disability, as well as the general decline in function associated with 'healthy' aging, including mental decline. Cognitive impairment as measured by standard tests of memory and concentration was found in one in six of 60–64-year-olds rising to one in four of 70–74-year-olds[4]. Severe memory impairment increases more dramatically with age (Figure 4.2),

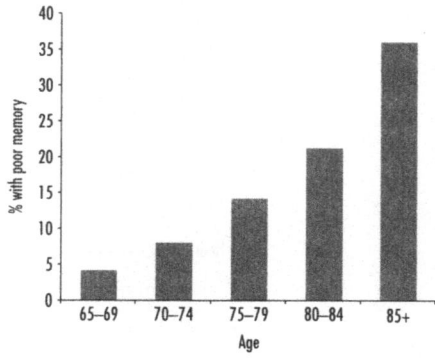

Figure 4.2 Percentage of US population with moderate or severe memory impairment (4 or fewer words recalled out of 20)[6].

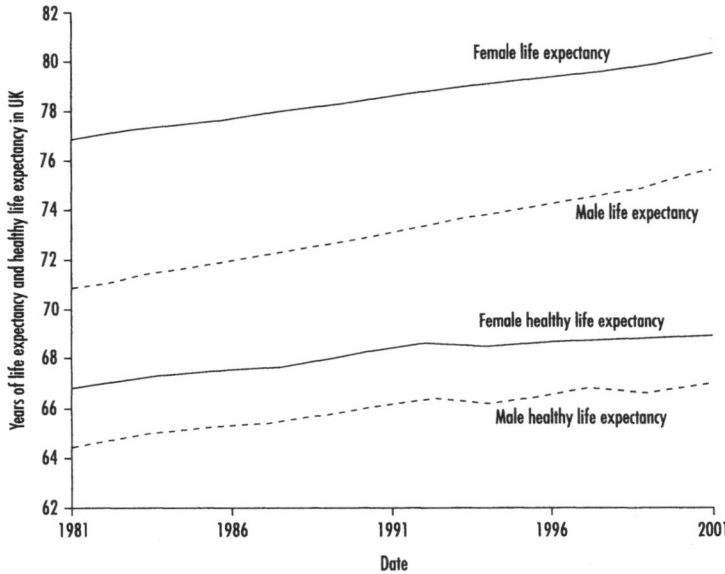

Figure 4.3 Life expectancy and healthy life expectancy at birth in UK. Office for National Statistics, UK.

so that some form of memory loss is almost universal over the age of 85 years.

The perception that increased life expectancy may bring with it a greater burden of disability towards the end of life led some researchers to calculate a new statistic of 'healthy life expectancy'. This shows that while life expectancy is steadily increasing, healthy life expectancy has more-or-less plateaued. In other words: unhealthy life expectancy increased (Figure 4.3).

The UK's Office for National Statistics stated in 2004:

The population of Great Britain has been living longer over the past 20 years, but the extra years have not necessarily been lived in good health. Life expectancy and healthy life expectancy (expected years of life in good or fairly good health) both increased between 1981 and 2001, with life expectancy increasing at a faster rate than healthy life expectancy. The difference between life expectancy and healthy life expectancy can be regarded as an estimate of the number of years a person can expect to live in poor health. In 1981 the expected time lived in poor health for males was 6.5 years. By 2001 this had risen to 8.7 years. Females can expect to live longer in poor health than males: in 1981 the expected time lived in poor health for females was 10.1 years, rising to 11.6 years in 2001.

Most of that ill health occurs at the end of life, so that we can currently expect about a decade of poor health before death, and that time is increasing rapidly (Figure 4.4). These are really sobering statistics: life expectancy in the UK rose by about 2.2 years between 1991 and 2001, but healthy life

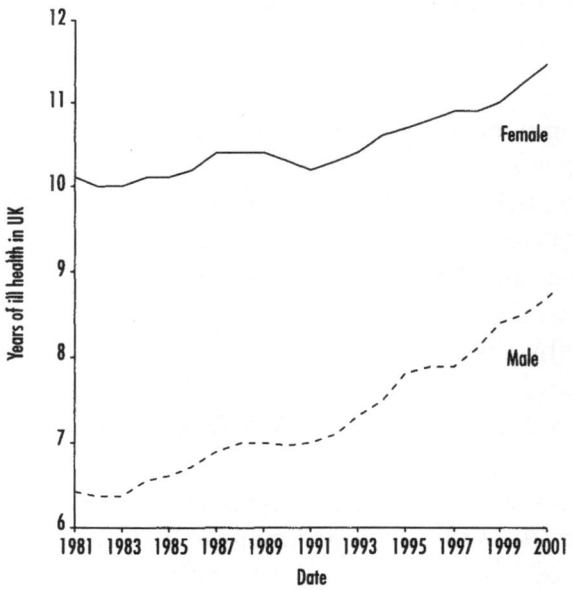

Figure 4.4 Unhealthy life expectancy in the UK (i.e. life expectancy – healthy life expectancy). Office for National Statistics, UK.

expectancy rose by only 0.6 years, while years in ill health rose by 1.6 years. Life is being extended, but it is being extended at the end rather than the beginning or middle, so most of that extra life is lived in ill health and disability. If we project these trends into the future we get the Tithonus scenario: immortality with infinite aging.

Dying is no fun. UK (General Household) surveys have looked at disability in the final year before death. In 1969, 30% of such people needed help getting in and out of the bath, dressing and undressing and/or washing for at least a year before death. By 1987 this had risen to 52%, indicating again that extended life is being bought at the price of increasing years of disability[5]. Mental confusion, depression and incontinence were all reported to occur over a longer time period before death in the 1987 survey than the 1969 survey, again probably due to the increased age at death. Depression generally increases with age, and 23% of non-institutionalized Americans over 85 are thought to be severely depressed[6]. This figure is probably higher in those who are institutionalized in hospitals, mental facilities, hospices and old-age care homes. Symptoms experienced in the last year of life were analyzed as a function of cause of death in a 1990 UK survey[7].

Those dying from cancer in general suffered from the most symptoms in their final year: 88% suffering pain, 54% breathlessness, 59% nausea and vomiting, 41% difficulty swallowing, 63% constipation, 41% mental confusion, 28% pressure sores, 40% urinary incontinence and 32% bowel incontinence. Those dying from heart disease or stroke in general suffered somewhat lower levels of these symptoms, but for a longer period before death, those dying from stroke suffered from higher levels of mental confusion and incontinence, and those dying from heart disease suffered higher levels of breathlessness; almost all suffered substantial pain.

Dying is also increasingly lonely. Of those dying of stroke, 67% died in hospital, 24% in a residential or nursing home, and only 9% at home. Fifty-nine per cent of dying women were widowed, and 44% had lived alone in the year prior to death. These figures were lower for those dying of cancer largely because they were younger at time of death. The 1996 GHS of private UK house-holds found 87% of people aged 65 and over living either alone or with only a spouse. This figure has been steadily increasing over time, while the proportion of elderly parents with at least one child living 10–15 minutes travelling distance away has been declining. This means that the elderly are increasingly isolated, and have reduced sources of informal help and care to draw on as they endure old age and approach death.

The proportion of elderly people living in the same house-hold as their children has been declining in every country that records these figures. In Denmark, Sweden and Holland the figure is now below one in ten. Living in nuclear families is increasingly becoming the norm, even in those countries such as Japan and India with a strong history of filial piety. Increas-ing levels of divorce are further fragmenting families. Increas-ingly, the elderly are left on their own. Over half the elderly in Denmark live on their own, and between 30 and 40% in most Western countries. Migration of the young into the cities in developing countries, such as China, leaves the old increas-ingly alone in the country. This change to elderly people living alone has occurred relatively recently: it was one in eight in the UK in 1945, and is now one in three, and the proportion is still rising rapidly.

In late middle and old age people are supposed to undergo a process of 'disengagement' from family, friends and society (Figure 4.5). This comes about for a variety of reasons. Retire-ment disengages the retiree from a circle of friends, col-leagues, and new people, from a longstanding and fixed routine, and from what has often become a purpose in life.

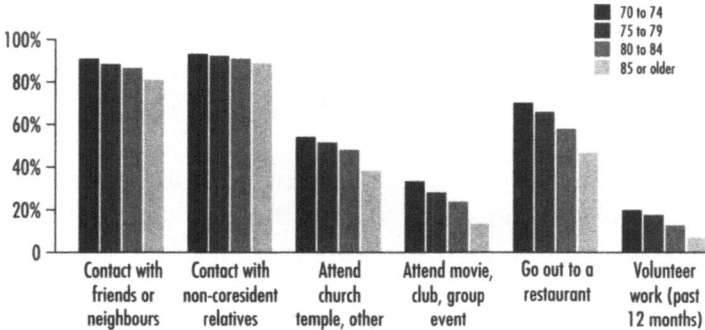

Figure 4.5 Percentage of (non-institutionalized) US citizens who report engaging in social activity in 1995. Data from a sample of persons asked whether they had engaged in the indicated activities in the last two weeks, or had done voluntary work in the preceding year. Source: Federal Interagency Forum on Aging-Related Statistics. *Older Americans 2000: Key Indicators of Well-Being.* Federal Interagency Forum on Aging-Related Statistics, Washington, DC, U.S. Government Printing Office. August 2000.

Children leaving home and the deaths of family and friends remove social support and identity, and a reason to care. Social and economic changes mean that most old people now live alone or with a spouse, but without children or grandchildren. Roughly 22% of men over 75 years lived alone in the US in 1998, while 52% of women similarly aged lived alone[6], reflecting the greater age to which women live.

Loss of health, mobility and energy removes the capacity to find new interests and friends (Figure 4.5). Loss of youth, the narrowing of possibilities and the contemplation of death can in themselves bring on a kind of mourning and bereavement. Social withdrawal and subsequent isolation may in some people be a protective strategy against further emotional loss, but old age often brings a paradoxical dependency on other people, which many hate. As Clive Seale (Sociology Professor at Brunel University in London) states:

The perception that useful life has finished is a more widespread experience when more survive to great old age, particularly in a society that has developed routinized methods for separating older people from the family, social and economic activities of younger groups.

Old people may become alienated from a modern society that they are less involved with and which values them less. Previously, close ties with an extended family and local community helped maintain their psychological wellbeing. Infirmity, repeated bereavements, acquired poverty and reduced opportunities for involvement are the burdens of old age. Tired of life, many may turn their faces to the wall and welcome death. Depression, delusions and anxiety disorders are common in the aged. Psychological stress, depression and bereavement all contribute to an increased susceptibility to disease and death. Of course, not all old people are depressed, many 'age successfully', but social and psychological death is common prior to physical death.

Towards the end

Julia Lawton was in her twenties when she started doing sociological research on 'the dying process' by working as a volunteer in a UK hospice for terminally ill patients for six months, during which time she experienced over 200 deaths. As she recorded in her diary, her first day was a bit of a shock:

I found my first real encounter with the patients in the wards slightly repugnant. One of the beds we stripped down was covered in excreta. The woman concerned has MS and suffers from incontinence.... A nurse was called into one of the cubicles to help out a woman who was vomiting.... A man walked up and down the corridor with his penis and catheter hanging out from his pyjamas. Another woman lay on the top of her bed revealing a bloated stomach and scab covered legs[8].

The modern hospice movement was started in the late 1960s in the UK by a small group of medics disillusioned with the medicalized care of dying patients in hospitals. They complained that dying patients were generally ignored by hospital staff, were not consulted, had inadequate pain control, and had little social, psychological or spiritual support. They were generally left alone, terrified, and in constant pain, or alternatively heavily drugged, until they died.

Cicely Saunders founded the first modern hospice, St Christopher's, in London in 1967, specifically to care for dying patients, with an emphasis on social and psychological support plus pain relief. The hospice movement subsequently spread through many Western countries, explicitly in opposition to established hospitals. However, economic pressures have caused the hospices to integrate with hospitals and diversify into day care and home care. One consequence, which Lawton chronicles, is that hospices have been increasingly reserved for patients with the worst medical symptoms, who benefit from symptom control but are less capable of benefiting from social and psychological support.

Lawton noted that few patients dying in the hospice had the capacity to prepare psychologically or spiritually for death. Many patients died either in extended sleep, in a coma, heavily sedated, or in extreme confusion or pain. Of those who remained conscious and able to communicate, most manifested an erosion of self. As one patient wryly suggested, it must be extremely difficult to engage terminal patients in conversation, as: 'After all, most of us in here have gone past the stage where we're into in depth conversations about the meaning of life and all that willy-nilly'.

Lawton reported a progressive erosion of self and disengagement from the world in dying patients. Many patients reported a progressive lapse into nothingness:

> For me the physical and mental are entwined. I've found as I've got weaker I've become a lot more apathetic and withdrawn.... I've abandoned a lot of my favourite pastimes. A couple of months ago I stopped doing the crossword in the newspaper. Last month I stopped reading the newspaper altogether. I've just lost interest. I suppose that's why so many patients here spend so much time sleeping. There's so few things we're able to do... so you just give up. (Hospice patient)[8]

Lawton suggested that as patients progressively lost control of their bodies and became dependent on other people and machines for basic functions, they moved from being subjects to being objects, even in their own regard. As many thinkers have pointed out, healthy people tend to take their bodies for granted and identify their selves with their minds, but ill people cannot take their bodies for granted, and their selves become bound up with the body. Lawton further suggested that some patients suffered a serious loss of self and personhood when their bodies became 'unbounded': for example, cancer patients with continual incontinence, vomiting, and/or emission of other bodily fluids as their bodies disintegrated, accompanied by repellant odours of excreta, vomit and rotting flesh. Such odours were a continual problem at the hospice.

All this casts some doubt on the idea that patients achieve personal growth and self-enhancement through an awareness of death and dying. This idea was developed particularly by the American psychiatrist Elizabeth Kübler-Ross in her book *On Death and Dying* (1969), which provided a major impetus for the development of the hospice movement in America. She proposed that in general people pass through five stages while dying: phase one is denial and disbelief, and this is followed by anger, bargaining and then depression, before finally reaching a peaceful state of acceptance.

Kübler-Ross originally applied these stages to any form of catastrophic personal loss, such as the death of a loved one or even divorce. She also claimed these steps do not necessarily come in order, nor are they all experienced by all patients, though she stated that a person will always experience at least two. In contrast Lawton found no evidence for these phases. Indeed, the patients were much more concerned with dying than death. 'The overwhelming impression I gained from patients was that their feelings were those of dulled resignation; of apathy, lethargy and exhaustion; of finally giving up'.

Because religious belief is so concerned with death and dying one might imagine that religious people experience death differently from others, but Lawton, Kübler-Ross and others have found no apparent difference in the experience of death between believers and non-believers. Lawton comments:

> Christine, threw her Bible out of her locker (quite literally), claiming that she could no longer believe in God after, 'seeing so much suffering around me' ... and she liken her experience to 'the nearest thing I can imagine to *hell on earth*'[8].

My own experience of death in a hospice was much more benign. My father, aged 80, died recently from pancreatic cancer. The acute illness lasted just three months, preceded by three months of digestive problems and moderate pain. The second operation to reconnect his pancreas was abandoned after a gallon of green tumourous goo was pumped out of his abdomen. My father, who had been a vet, recognized this goo as a sign of utter hopelessness, and while lying on the operating table advised the surgeon to stop wasting their time – a request to which the surgeon gratefully assented. My father had himself released from hospital the next day, to go home to die. He spent three brave weeks struggling to live his life as normally as possible to the end. Then, when the pain and nausea were too much, he spent another three weeks in a hospice. His skin by this time was completely green and his eyes yellow. He lost most of his weight and became a bag of bones, which made movement and sitting hard. But his mind remained clear as a bell, clear enough to see the ruin of his body, and clear enough to see beyond it. The hospice got on top of his pain, and was generally brilliant.

We spent two days together going through his life in detail, more for my benefit than his. We discussed his death. I tried to source some barbiturates to help him die if necessary. But even with international pharmacies online it is ridiculously difficult to get hold of barbiturates or anything else to kill yourself without considerable pain or mess. I returned one last time to the hospice on a vile rainy day – I hadn't seen him for a week – and it was like something from a horror movie – I had never seen someone so dead who was still alive. I daren't make any sound in case he somehow woke and he and I were confronted with his existing in such a degraded state. I sat listening to his halting breathing for hours. The darkness merged with the silence and I crept out. He died in the night.

Death with dignity is rare. As the doctor Sherwin Nuland, notes:

> The quest to achieve true dignity fails when our bodies fail. Occasionally – very occasionally – unique circumstances of death will be granted to someone with a unique personality, and that lucky combination will make it happen, but

such a confluence of fortune is uncommon, and, in any case, not to be expected by any but a very few people.... The good death has increasingly become a myth.

Most people (70% in the US) would prefer to die at home. However, most people (75% in the US) end up dying somewhere else, such as a hospital or nursing home. Where people die has changed dramatically over the last 100 years. Over much of the twentieth century there was a steady rise in the proportion dying in hospital rather than at home, rising to a peak of 70% in the early 1980s in the US. Since then there has been a rapid shift away from hospitals as the site of death to home and nursing homes, so that in 2000 roughly 50% of US citizens died in hospital, 25% at home and 25% in nursing homes. Less than 1% of US deaths occurred in hospices[3].

The final stages of the average American death occur in a hospital's intensive care unit, a gleaming showcase of modern technological medicine, bristling with polished chrome machines and flashing computer screens. This is the theatre of aggressively interventionist medicine, which is also massively expensive. Frail, withered bodies are plugged into virile machines via a forest of tubes and wires. As the body starts to pack in, individual organs or body functions fail. Outside the hospital this would rapidly lead to death, but in the ICU the organ's function may be replaced by a machine. As death unfolds other organs fail, and may be replaced by other machines. It is almost as if the machines enable the person to defy the gravity of death. But this gravity can only be defied for so long.

There comes a stage in all deaths where respiration and circulation fail, a stage that was traditionally defined as the irreversible point of death. Nowadays this is the cue for the specialist resuscitation teams to attempt resuscitation, via heart massage, defibrillators and insulin therapy. Keeping people alive in intensive care units is massively expensive in terms of money, machinery, time and effort. Often the dying person's machine is turned off, leading to 'death', simply because the resources have run out, whether that is the availability of intensive care beds or the doctor's interest. As the intensive care doctor Professor Mervyn Singer at University College Hospital in London has pointed out, to the embarrassment of his colleagues, most people finally die in intensive care units not because the body gives up, but because the medics get bored of trying and/or decide to conserve resources.

I toured an intensive care unit in London recently, and I was struck by what an awful place it would be to die. It is an exciting place to live and work because of the intense dramas of life and death being played out there on open display. People were coming in and out on the edge of death; relatives were pacing, crying, weeping and grieving; and doctors and nurses were running, shouting, consulting and joking. All the doctors and nurses I talked to expressed a strong preference for themselves to die at home, rather than in an intensive care unit. Unfortunately, in the US, partly because of the fear of litigation, when someone is in danger of dying in hospital they are sent to the intensive care unit. Since the primary function of these units is to prevent death, they are crappy places to die.

By 2020 it is projected that 40% of deaths will occur in nursing homes in the US. The shift to nursing homes partly reflects the aging population and the chronic disability of that population prior to death. Nearly one in two US citizens who live into their 80s will spend time in a nursing home. Of those in a US nursing home, nearly one in two had some level of pain that may impair mobility, social relationships, sleep and/or self worth. Of those persons who were terminally ill, 45% had moderate daily or excruciating pain. Pain is the symptom most expected and most feared by dying patients. Unrelieved pain can have enormous physiological and psychological effects on patients and their loved ones.

The decline in function prior to death takes different trajectories depending on cause of death. Cancer patients normally have little functional impairment six months prior to death, then undergo a rapid and deep decline in functions such as moving, eating, dressing, bathing and using the toilet. Deaths due to heart disease and diabetes are preceded by a much slower decline in function extending over years, so that for example 12 months before this type of death 30–40% of people have trouble getting into or out of bed (Figure 4.6). Apart from these functional disabilities, fatal illness is often accompanied by the following symptoms: pain, shortness of breath, digestive problems, incontinence, bedsores, fatigue, depression and anxiety, confusion and finally unconsciousness.

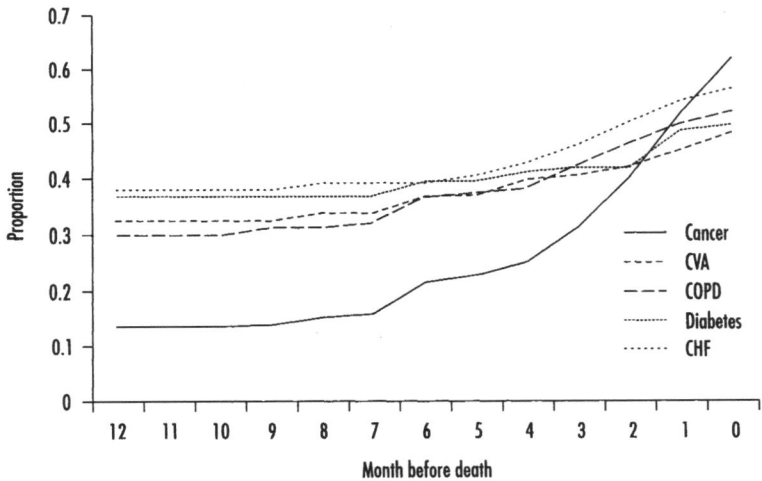

Figure 4.6 Proportion of people (with different diseases) having trouble getting into or out of a chair in the year preceding death. http://www.chcr.brown.edu/dying/.

To palliate these symptoms, the dying are usually given increasing doses of opiates as they approach death, and therefore are usually unconscious or only partly conscious in the days and hours before the final event, when the lungs and heart stop functioning. Because the dying often lose their ability to swallow, the final event is often preceded by the death rattle: a gurgling or rattle-like noise produced by the accumulation of excessive respiratory secretions in the throat. In the final event the heart will stop, with the consequence that energy is no longer delivered to any organ, resulting in irreversible damage to the brain within 15 minutes, irreversible damage to the heart itself in 30 min, and subsequent irreversible damage everywhere else.

interlude 4
the immortals of luggnagg

Sometimes, though very rarely, a Child happened to be born in a Family with a red circular Spot in the Forehead, directly over the left Eye-brow, which was an infallible mark that it should never dye[1].

Thus is Captain Gulliver introduced to the Immortals of the Kingdom of Luggnagg. He immediately enthuses and waxes lyrical on the great advantages he would accrue were he too to be blessed with immortality. But the pitying Luggnaggian disabuses him of his naivety, and...

After this Preface, he gave me a particular Account of the *Struldbrugs* among them. He said they commonly acted like Mortals, till about Thirty Years old, after which by Degrees they grew melancholy and dejected, increasing in both till they came to Four-score. This he learned from their own Confession; for otherwise there not being above two or three of that Species born in an Age, they were too few to form a general Observation by. When they came to Fourscore Years, which is reckoned the Extremity of living in this Country, they had not only all the Follies and Infirmities of other old Men, but many more which arose from the dreadful Prospect of never dying. They were not only opinionative, peevish, covetous, morose, vain, talkative; but uncapable of Friendship, and dead to all natural Affection, which never descended below their Grand-children. Envy and impotent Desires, are their prevailing Passions. But those Objects against which their Envy seems principally directed, are the Vices of the younger Sort, and the Deaths of the old. By reflecting on the former, they find themselves cut off from all Possibility of Pleasure; and whenever they see a Funeral, they lament and repine that others have gone to an Harbour of Rest, to which they themselves never can hope to arrive. They have no

Remembrance of any thing but what they learned and observed in their Youth and middle Age, and even that is very imperfect: And for the Truth or Particulars of any Fact, it is safer to depend on common Traditions than upon their best Recollections. The least miserable among them, appear to be those who turn to Dotage, and entirely lose their Memories; these meet with more Pity and Assistance, because they want many bad Qualities which abound in others[1].

We may laugh now, but will we be crying later? Jonathan Swift, the author of *Gulliver's Travels*, was a master of satire and a consummate explorer of the human condition. His account of the Struldbrugs' fate is a prescient description of where we are headed in our own voyage into the nether regions of extreme old age. The aged Swift himself, as he had feared, succumbed to dementia, and was declared of 'unsound mind and memory' three years before he died in 1745 at the age of 77. He managed to keep some of his sense of humour, though. His last will and testament provided funds to establish a mental hospital around Dublin for 'ideots & lunaticks' because 'No Nation wanted [needed] it so much'.

5

losing your marbles

Brain aging

Physical aging might be bearable if we could hang on to our minds. But can we? It is tempting to think that peripheral things are lost with age, but the core of self is preserved. We all know old people who, while having poor memory and slower responses, are as 'sharp as a pin' and 'all there'. Most of us can also think of very old people who are *not* as sharp as a pin and are *not* all there. Mental aging hits different people at different rates and in different ways. However, loss of memory and loss of the ability to form new memories is almost universal with age (Figure 4.2). Also, the senses of sight, hearing, taste and smell are dulled in everyone, and reaction times slow. Linguistic ability deteriorates; IQ declines; creative thinking falls off; mental productivity wanes. And that's all before we consider the chronic illness, neurodegenerative diseases, depression, anxiety and social-psychological problems of the aged! None of this is good news, and there is no point pretending otherwise: it is one of the worst horrors of the human condition.

One of the most obvious changes with age is sensory loss. From the age of about 25 hearing loss begins, and from the age of about 35 people start to complain about it. Perception of higher frequencies declines continuously. Other sounds appear more rapid and loud. This decline is more-or-less universal, but impairment in the capacity to appreciate music and the spoken word is also common, which can impact on the ability to hold a conversation and make friends. Visual acuity also declines with age. Clouding of the lens in the eye starts at age 20, reduced ability to focus is a well-known characteristic of aging, and loss of visual processing in eye and brain commences later. Ability to adjust to different light levels declines particularly after sixty. One quarter of those over 60 have some degree of macular degeneration of the eyes. Colour

vision declines, making it more difficult to distinguish between blues and greens. All these visual changes may impact
on the ability to drive. The senses of touch, smell, taste and
balance also decline with age, making the world a less joyous
place to be.

Mental speed declines with age as measured by reaction
time, and the slowing is more marked for more complex
tests requiring choice. Decreased nerve conduction speed,
impaired sensory functions and slowed muscle movements
may all contribute. But slowing of the mechanism transferring
signals from neuron to neuron is thought to be the most
important cause. Brain speed can also be measured as the
response time of the brain's electrical activity to a visual, audi-
tory or tactile stimulus. Typically an electrical response can be
measured in the brain 300 milliseconds after the stimulus. This
slows to 400 milliseconds at age 80, and becomes much
slower in demented patients, indicating a slowing in neuronal
conduction, communication and processing.

Neurons in the brain die throughout life. Up to a certain
point the loss of neurons is compensated by the birth of
new ones. That point occurs early: at about three years of
age. After that there is a long slow decline, accelerating
towards old age. And in old age the brain slowly shrinks.
Loss of neurons is faster in some areas than others: the frontal lobes for example. These are massively expanded in
humans compared to other animals, and are apparently
what make us 'human'. They are thought to be responsible
for higher mental skills, such as planning and judgement,
as well as coordinating emotion, attention and memory.
During every decade of old age the frontal lobes shrink by
5%. Neuronal loss here causes the decline of mental abilities
with age[1].

This is normal brain aging, and it is more or less unavoidable.

Alzheimer's and other dementias

If one word strikes more terror into the heart of the aged than death, it is 'Alzheimer's'. Alzheimer's disease is the main cause of dementia or mental degeneration. To depend on others for everything: spoonfuls of food, the toilet, a change of soiled clothes – is intolerable even to imagine. But to slowly lose your wits, your personality, your self, and end up babbling inanities is a cruel fate indeed.

Alzheimer's currently strikes 11% of the US population over the age of 65, and that proportion is rising. The total number of Americans with the disease is estimated to be about 2 million in 1980, 4.5 million in 2002, and between 11 and 16 million in 2050[2] (Figure 2.5). Total dementia in the UK is estimated as 0.7 million in 2002, and predicted to be 1.7 in 2050, of which two thirds is due to Alzheimer's[3]. The main reason for this growing plague is the huge increase in the number of people surviving to old and very old age (Figure 2.4). Increasing age is the greatest risk factor for Alzheimer's: prevalence of the disease is about 1% at 65 years of age, and doubles every five years thereafter, so that about 25% of 85-year-olds are thought to have Alzheimer's[4]. This is very frightening. It means that according to current predictions someone born in the twenty-first century has a one in three chance of getting Alzheimer's disease or some other form of dementia before they die. This more-or-less guarantees a 'living death', even if we avoid all the other cognitive pitfalls of age.

From the time of diagnosis, people with Alzheimer's disease survive about half as long as those of similar age without dementia. Alzheimer's is currently the eighth most common cause of death in the US, and rising. Even so, a person with Alzheimer's disease will live an average of eight years and as many as 20 years or more from the onset of symptoms as estimated by relatives.

The history of Alzheimer's is illustrative of the difficulties and pitfalls of identifying a new disease. The cause, diagnosis, and even the existence of Alzheimer's are still controversial. Identifying a new disease is not easy: many have been identified that do not exist; some have a spectrum that overlaps with other diseases. Most new diseases start as a collection of symptoms with no known cause; many end up as a collection of symptoms with no known cause – only now they have a name. Senility, meaning a loss of mental faculties in the aged, has been recognized for thousands of years. But the term dementia was not introduced until 1801, to distinguish a specific form of clinical madness characterized by 'incoherence'.

The inventor/discoverer of this distinction was Philippe Pinel, senior physician of Le Salpêtrière, the Paris asylum for incurably ill and insane women. When Pinel was appointed he introduced the novel principle of the 'moral treatment of insanity'. Insane women who had been chained for 30 or 40 years were unchained. Discarding the long-popular equation of mental illness with demoniacal possession, Pinel regarded mental illness as the result of excessive social and psychological stresses. He did away with treatments such as bleeding, purging and blistering, and introduced a regimen that included friendly contact with the patient, discussion of personal difficulties, and a program of purposeful activities. In *Nosographie Philosophique* (1798; *Philosophical Classification of Diseases*) Pinel made one of the first attempts to distinguish between different forms of mental symptoms, illness and disease. And in 1801 he first described '*démence*' ('dementia' in English) as a kind of incoherence of the mental faculties, characterized as follows:

Rapid succession or uninterrupted alternation of isolated ideas, and evanescent and unconnected emotions. Continually repeated acts of extravagance; complete forgetfulness of every previous state; diminished sensibility to external impressions; abolition of the faculty of judgment; perpetual activity.

This is still a reasonably good description of dementia of the Alzheimer's type, although a modern definition would emphasize the progressive deterioration of memory, personality and intellect. Subsequent work at La Salpêtrière distinguished between 'presenile dementia', first occurring in middle age, and 'senile dementia' or general 'senility', which was considered to affect those over 65 years. Autopsies on demented patients showed that the brain was often shrunken due to the loss of about 30% of the neurons. But it wasn't until 1906 that Alois Alzheimer, a German neuropathologist, used the microscope on the brain to distinguish different types of dementia in a meaningful way. In the autopsy of a 55-year-old person who died with severe dementia, he noted the presence in the brain of two abnormalities: neuritic plaques and neurofibrillary tangles.

Neuritic plaques consist of bits of dead neurons surrounded by a sticky protein called β-amyloid. Neurofibrillary tangles are twisted fibres, located within nerve cells, made from a protein called tau. Whether these plaques and tangles are the cause or consequence of the disease is still unclear, although there is now mounting evidence that they do indeed kill the neurons. Plaques and tangles are present in the brains of most healthy people, although at much lower levels than seen in Alzheimer's patients, so there may be some continuity between the disease and normal brain aging. Because the disease was often seen in association with arteriosclerosis of brain blood vessels, it was thought to be caused by this blockage of the brain's blood supply. And because Alzheimer's happened to be first noted in a relatively young person, it was long regarded as a form of presenile dementia. Thus for 60 years after its discovery the disease was regarded as a relatively rare form of dementia starting in middle age.

Alzheimer's disease was rescued from its obscurity in the 1960s by Sir Martin Roth at the University of Newcastle in the UK. Roth was the leading British psychiatrist of his day, and shepherded psychiatry from the age of Freud to the era of Prozac. He and his colleagues found that the plaques and tangles thought to be characteristic of presenile dementia were just as prevalent in senile dementia, which was vastly more common. A London conference in 1968 caused a sea change in the classification of dementia: from then on, dementia was increasingly diagnosed as of the Alzheimer's type. Suddenly Alzheimer's disease was everywhere. Patients and relatives were demanding action, medics and scientists were demanding money, and governments and agencies began supplying it in spades. The Alzheimer's research budget in the US grew 800-fold during the 1980s. The National Institute of Aging (NIA) was founded in the USA to combat Alzheimer's. According to the World Health Organization this previously exotic and obscure disease was now one of the leading scourges of humanity. The total cost of Alzheimer's to the US economy is currently estimated at $100 billion per year, and rising.

New research in the 1990s began to focus on the β-amyloid plaques as the source of the disease. This was largely as a result of the discovery that mutations in the gene for the β-amyloid protein or related proteins could cause the disease. These mutations were responsible for only a very small percentage of Alzheimer's patients, but it demonstrated the principle that dysfunction of this protein could cause the disease. Subsequently it was shown that adding β-amyloid protein to isolated neurons could kill them, but it was still unclear whether the protein contributes to killing neurons in the intact brain. To test this, mice were genetically engineered to overexpress (that is to produce more than the usual amount of) the β-amyloid protein in their brains, and these mice were found to develop a neurodegenerative disease which could be thought of as an Alzheimer's disease of mice.

The key question remained: would removal of this protein prevent the disease? To test this, a vaccine was developed against β-amyloid – basically a piece of the protein was injected into the blood causing the body to produce antibodies against the protein, resulting in its removal from blood and other body tissues. Use of this vaccine in mice that had been genetically engineered to overproduce the β-amyloid protein resulted in spectacularly successful clearance of the protein from the mouse brains. In 2002 the medical community all over the world held its breath when the first clinical trial of this vaccine was attempted in human Alzheimer's patients. Disaster struck when 15 of the treated patients developed life-threatening inflammation in the brain. But there is still hope that this general vaccination approach may eventually contribute to treatment.

As the millennium arrived, the pendulum started to swing back on Alzheimer's. The old idea that the disease is due to arteriosclerosis in the brain has been resurrected. It is now thought that Alzheimer's may overlap with vascular dementia in various ways. And there may be some overlap with normal brain aging: both may be promoted by free radicals (reactive oxygen species) and inflammation. Inflammation and free radicals are normally produced in response to some other stimulus or damage, so the real, ultimate cause of Alzheimer's remains unclear.

As we have seen, vascular dementia was also first described by Alois Alzheimer, when he saw multiple small regions of dead tissue in the brains of demented patients, suggesting that the patient had suffered from a series of mini-strokes. Vascular dementia is dementia of vascular origin, i.e. caused by chronic or intermittent disruption of the blood supply within the brain. Indeed, vascular dementia is often accompanied by arteriosclerosis and normal strokes. There are different types of vascular dementia and its diagnosis is not easy and is currently in flux. Vascular dementia was briefly eclipsed by Alzheimer's in the attention of the medical community, but is now coming back strongly, with anything from 10–70% of all dementias now being described as vascular in origin.

Alzheimer's predominates in North America and Northern Europe (where vascular dementia accounts for about 20% of all dementia), but vascular dementia dominates in Japan and black Africa. Whether this difference results from different genetics or simply different diagnostics is unclear. Generally we are diagnosed with Alzheimer's if the dementia is slow in onset and progress, while we are diagnosed with vascular dementia if it is rapid or we have previously had a stroke. But these two forms of dementia overlap.

In 1993 a more tangible connection between Alzheimer's and vascular disease was discovered. Gene hunters found a gene known as APOE on chromosome 19, various versions of which affected the incidence of both Alzheimer's and arteriosclerosis. If you have a particular version of this gene, and many people do, you are in for some serious trouble. People may have slightly different versions of this gene, differing by one letter (nucleotide) of the DNA sequence. The three common versions are known as APOE2, APOE3 and APOE4. Since almost every gene in our genomes is present in two copies, one on the paternal chromosome and one on the maternal chromosome, we have a pair of APOE genes. Thus in Europe, for example, 39% of people have two E3 genes, 7% have two E4 genes, 4% have two E2 genes, and the rest have combinations of these three versions. Roughly 30% of Europeans (and higher elsewhere) have at least one E4 gene, and these people have a substantially increased risk of both arteriosclerosis and Alzheimer's. For example, having one E4 gene roughly doubles your chance of getting Alzheimer's, while having two copies more than quadruples your chances. Most people with two E4 genes will end up with Alzheimer's disease, and its onset will be at a relatively early age.

E4 differs from E3 by a single letter. How does this tiny difference make you dramatically more susceptible to both arteriosclerosis and Alzheimer's? Unfortunately we don't really know – otherwise we might be able to do something about it. We do know that the APOE protein ferries fats around in the blood, and enables cells to take up those fats. APOE4 is less good at its job, so less fat gets taken up from the blood, potentially allowing more fat to build up in the arteries resulting in arteriosclerosis. Arteriosclerosis in brain arteries may in turn contribute to Alzheimer's. But the story might be more complicated than that. APOE4 can promote free radical production, which may in turn promote both Alzheimer's and arteriosclerosis.

A third type of dementia has erupted recently, known as Lewy body disease (or dementia with Lewy bodies). The symptoms are intermediate between Alzheimer's and Parkinson's disease. The distinguishing feature of Lewy body disease is the presence of small, round aggregates within the neurons. Lewy bodies are also found in the brains of Parkinson's disease patients. In this case they are restricted to the dying neurons of the brain region that controls body movements, the substantia nigra. Parkinson's disease thus mainly affects body movements, whereas Lewy body disease affects many different brain functions.

We don't know what causes either of these diseases, but current research suggests that it has something to do with the aggregates that make up the Lewy bodies. They are made up of a normal protein called synuclein, which has been damaged, possibly by free radicals. The damaged synuclein then clumps together to form fibres, which then aggregate to form the round Lewy bodies that injure the cell and cannot be broken down. When the cell finally gives up and dies, the toxic aggregates are released and may then spread to other cells, or activate inflammation that damages surrounding cells.

The slow accumulation with age of indigestible aggregates appears to be a common theme also in other degenerative diseases: β-amyloid and other proteins aggregate as plaques outside neurons in Alzheimer's disease; and a mutated protein, Huntingtin, sticks to itself and aggregates inside neurons in Huntington's disease.

Parkinson's disease does not normally cause dementia, at least in the early stages (although 20–40% develop dementia in late stages). Rather, it gradually paralyses the patient, without affecting the mind. The prevalence of Parkinson's also increases dramatically with age, and therefore is predicted to increase in the future. There are currently thought to be 1.2 million Parkinson's patients in the US, and this number is increasing by about 150,000 per decade. Parkinson's has recently increased to be the 13th most common cause of death in the UK. PD is not by itself a fatal disease, but in the late stages it may cause complications such as choking, pneumonia, and falls that can lead to death.

One famous Parkinson's sufferer was Pope John Paul II, and his relatively public end illustrates the multifaceted nature of death today. The pope finally died on 2 April 2005 at the age of 84, having had Parkinson's disease for about a decade. His death involved a slowly accumulating constellation of dysfunctions, including cancer and heart disease. Starting in about 1992, John Paul II's health slowly declined. He began to suffer from an increasingly slurred speech, difficulty in hearing and greatly reduced mobility, probably marking the onset of Parkinson's; he also had hip and knee ailments. He used a wheelchair and 'Popemobile'. The contrast between the athletic John Paul of the 1970s and the declining John Paul of later years was striking.

In 2000 he apparently seriously considered resigning the papacy due to failing health. For the last few years he had difficulty in hearing and saying more than a few sentences, and he had severe arthritis. He remained mentally alert, but with a failing memory. He also had hypertension, ischaemic heart disease and difficulty breathing. The final event of his death was preceded by months of ill health, including several bouts of respiratory infection requiring hospitalization, a tracheotomy (a tube inserted into his windpipe) to relieve breathing difficulties, and a feeding tube.

These symptoms would have been caused or exacerbated by the Parkinson's disease, which causes difficulty in coughing, breathing and swallowing. He had a worsening heart condition; then his kidneys started to fail, causing a urinary tract infection that resulted in high fever, intermittent loss of consciousness and blood poisoning, which reduced his blood pressure and put further pressure on his failing heart, until he died in a coma with multiple organ failure.

Our brains are powered by mitochondria, electricity generators squirming around in our cells[5]. Mitochondria have their own DNA, and this has been found to be mutated in the brains of Alzheimer's and Parkinson's patients[6,7]. These mutations result in the cells producing less energy and possibly more free radicals, potentially leading to neuronal death. Frighteningly, a number of environmental toxins have been found to attack mitochondria and cause Parkinson's disease. In the 1980s hundreds of young adults in California appeared to develop Parkinson's almost overnight. It turned out that a toxic contaminant of a designer drug that they were abusing was attacking their brain mitochondria, causing the disease. More recently it was found that a common pesticide, rotenone, well known to be a potent mitochondrial toxin, could also induce Parkinson's, at least in animals. Thus mitochondrial dysfunction, prompted by genetics, environment or aging, may kill neurons and contribute to dementia.

Traumatic brain injury (TBI) is caused by an external blow to the head rather than internal degeneration, and it is the leading cause of death and disability in children and young adults, mainly due to transport accidents. However, people over the age of 75 have the highest rate of TBI, largely due to falls. There are estimated to be about 5.8 million TBI patients currently in the US, higher than Alzheimer's and Parkinson's combined. And, you guessed it, this figure is rising due to the aging population. TBI also greatly increases your risk of subsequently developing Alzheimer's and Parkinson's. Watch your feet!

Vascular dementia, Alzheimer's, Lewy body and Parkinson's are all neurodegenerative diseases that emerged in the twentieth century, together with many others including Pick's, Huntington's, multiple sclerosis, motor neuron diseases, and fronto-temporal dementias. These age-related diseases became apparent in the last hundred years, not because they had not existed before, but because many more people survived to old age. As life expectancy continues to increase, the incidence of neurodegeneration and dementia will continue to increase, unless we can discover a cure for these diseases or to aging itself. This is not likely to happen any time soon.

Normal brain aging, in the absence of neurodegenerative disease, is accompanied by many of the same changes but in a milder form. Neurons are lost, but fewer and more slowly. Insoluble deposits of β-amyloid and Lewy bodies accumulate, but not as dramatically. The aged brain is characterized by a 'smoldering' low-level inflammation, producing damaging free radicals. Mitochondrial DNA mutations build up in the brain with age, but to a lower level than with disease, and energy generation within the brain is moderately impaired.

It is beginning to seem as if normal brain aging is itself a form of neurodegenerative disease, but milder and slower. Recent evidence suggests there may be a continuum between normal brain aging and dementia in terms of the changes in the brain[8]. That is, the same pathological processes are occurring in all of us, but faster in some than in others, such that they are pushed over some threshold precipitating disease at an earlier age. If true, this continuum model suggests that 'normal' brain aging can be regarded as a disease itself, and that if we live long enough we will all get dementia.

On the other hand, German neurologist Heiko Braak suggests that normal brain aging is not a separate disease but rather an early stage of Alzheimer's, Parkinson's or some other neurodegenerative disease. Braak showed that Parkinson's disease actually starts in a small area of the brain, and then slowly expands into neighbouring areas. The expansion is suggestive of an infectious disease. The damage spreads over a significant area of the brain before the patient actually notices that anything is wrong. Braak came to a similar conclusion in relation to Alzheimer's disease. It starts in a specific area of the brain, and spreads to other areas in stages, but the early stages have no obvious symptoms. Hence we have the idea that 'normal' brain aging may actually be due to a neurodegenerative disease at a relatively mild stage before it would be diagnosed[9].

Dementia and normal brain aging erode the self, and prevent it being unitary and unchanging. Someone with Alzheimer's cannot have a united or unchanging self, and may end up with no self at all to speak of. Normal brain aging may head in the same direction, though at a less dramatic rate.

Becoming demented is perhaps one of the worst fates that nature subjects us to. But is being demented, even brain aged, necessarily a bad thing? It's obviously a bad thing for society and for the people close to the 'patient', but after a certain level of loss of memory, identity and reasoning, the sufferer may not care, any more than a baby cares about its own lack of abilities. While people entering into and passing through Alzheimer's experience extreme depression, those with severe Alzheimer's are reported to be no less happy than the rest of us. Perhaps we should not fear having dementia, anymore than we should fear being dead – it is becoming demented and becoming dead that we should fear. As the science fiction writer Isaac Asimov said: 'Life is pleasant, death is peaceful. It's the transition that is troublesome'.

interlude 5
the mortal soul

I remember Sunday afternoons when I was a child – long inter-
minable afternoons – it always seemed to be raining, and
there was never anything to do. Time weighing heavily, we
used to watch television disconsolately; usually an ancient B-
movie filled the yawning hours. Many of those films were
instantly forgettable, while many formed the fabric of my day-
dreams and nightmares for the rest of my life. One left me
with a gnawing, aching sense of unease. There was a group of
adults who are clearly mentally retarded. They are playing
happily in the sunshine, like children, but there is a sense of
the unease that we might have amongst such people. These
adults share with children a lack of self-awareness: awareness
either of their condition or of themselves. The camera focuses
on one man alone on a swing: he is in his thirties, he is rocking
back and forth, he is mumbling a nursery rhyme to himself
and his face is rapidly alternating between a frown and a
smile. This man is to be our hero.

The story cuts to a Frankensteinish scientific laboratory, and
a clean-cut scientist with an ecstatic smile holds up a test tube
full of steaming, green gunk. 'I have done it' he thunders
triumphantly to his suitably awed assistants. After further
appropriate twists-and-turns of the plot, the green gunk is ad-
ministered to our mentally retarded hero, and the assembled
scientists stand back. Slowly (it takes a few weeks) the hero
emerges from his retarded state into the full self-awareness of
adulthood. He becomes aware: he becomes aware of himself
and others and he becomes aware that he was retarded and is
no longer.

But his growth of awareness, intelligence and self-consciousness do not stop at the normal adult level: the green gunk
propels him on to previously unknown levels of super-con-
sciousness – so far beyond our own, normal, adult levels of
intelligence, that we appear as mere children or retarded
adults to our hero. Everything he see, hears or feels is super-

charged with reams of streaming interpretation, explanations, associations, connections. Everything in the world is super-packed with a kaleidoscope of multidimensional meanings, choreographed exactly with everything else, so that every-thing is inevitable and understandable to him, who has be-come almost a god. And the most crystal-clear entity in this new world of his is himself: he understands every nuance of his being, every petty anxiety he feels resonates with its subtlest personal and evolutionary cause. His self-consciousness has reached an almost painful pitch. But things cannot stay as they are.

The story cuts to an excited scientific congress. The green gunk scientist stands on the stage and points the attention of the assembled professors to a projected image of a retarded adult rocking back and forth on a swing, mumbling to himself. He points to the green gunk, and then triumphantly draws aside a curtain to present our hero. He stands alone, eerily motionless in the centre of the stage as the projected image of his former self flickers across him, and intently studies his audi-ence, slowly becoming animated with excited whispers. The intensity of his eyes, the knowingness of his frown and the sad-ness of his faint smile are enough to inform his audience that a dramatic transformation has occurred. But now our hero calls for silence.

He tells them the biochemical, electrophysiological and the psychological origins of his former state and its transformation in terms that make the professors' heads spin. There is thun-derous applause, and much backslapping. He then pauses, and tells them he has a personal announcement to make. He tells them his analysis has shown that the green-gunk effect is entirely temporary: his brain will inevitably return to its former retarded state, he will inevitable sink into senility and there is nothing that he or anyone else can do about it. He stops and slowly walks off stage to the accompaniment of a stunned silence, the film of his former self still flickering on the screen, rocking back and forward.

In the months ahead he becomes a super-scientist and throws himself into trying every possible avenue to devise a new super green gunk to arrest his demise. But even before he starts he knows it is doomed, and he is all too aware of his failing powers of awareness, his slow deterioration of intellect, and slipping self-consciousness. Small slips of concentration fill him with a dull dread, an inexorable force is dragging him towards his inevitable fate: to pick him apart and leave him an idiot. He is being shorn of all the insights and subtle feelings so painfully gained. All his heaviness and depth is leaving. He is becoming as light and shallow and empty as a child, playing alone upon a beach. The film ends as he sits down upon a swing; a fleeting expression of horror is replaced by blankness as he rocks back and forth.

This film is a parable of life and death, or rather of growing into life and growing out of life into death. Life would be very different if we were to be born adult with full consciousness and intelligence, instead of growing into life almost impercep-tibly, such that the enormity of being alive is almost invisible. Hardly have we realized that we are alive when we realize that we must die. The cruel indignity of death seems almost per-verse. Why make us spend long years struggling to learn to walk and talk, to do things properly and have pride in our-selves? Why, before our own eyes and those of our children, does this cruel fate slowly strip us naked, take our teeth and hair, make us babble and froth at the mouth, hack pieces out of us, and rob us of our senses?

The story also suggests that the self is plastic: it can change with age, it can change with experience, and it can be radi-cally different in different people at different times. The film illustrates the analogue natures of death and self: we can be partly here and partly not. Why have the analogue natures of death and self been so fiercely resisted in Western society?

Western society encourages us to believe that the core of self is stable throughout life, and that, although each individual is unique and separate, we are in essence similar to most other people in the world. We believe that it is possible to empathize with any other human, to project ourselves into their world. This idea of equality of selves is partly derived from the concept of soul in Western religions: that there is a core essence to us that is unchanging throughout life (and death) and that is similar in different people. This contrasts with Eastern religions that emphasize differences between people, the evolution of souls, and even encourage the disintegration of self. In Western society the self is supreme, but all selves are supposedly equal (although it is not always possible for individuals to hold both values simultaneously). The liberation of slaves and women, the disintegration of royalty and class, the triumph of democracy and human rights are all founded on the ideas of 'liberty, equality and fraternity', and are the welcome fruits of Enlightenment thinking. It is hardly surprising then that we resist the idea that the self has parts, that it is different in different people, and that it changes throughout life.

Do we really want to regard old or very old people as half-alive, as half-selves, as half-human? This would appear disastrous both for social attitudes to the aged and for personal attitudes to our own aged future. We are scared of this concept both for ourselves and for other people. We don't want to become half-people, we don't want others to think of us as half-people, and we don't want to think of ourselves, our parents or others as half-people. But that is partly because we don't want to think about the reality at all – the reality of dementia for example – we want to shut it out, we want to live forever and ever in never-never land. We would rather imagine that everything is fine and dandy, that everyone in extreme old age is wise, jovial and rosy-cheeked. But what is the consequence of that collective myopia? Everyone who does not fit our vision of successful aging is banished to an institution, banished from the television, banished from our collective vision. We do not want to see or speak to disabled, depressed or demented old people; we would rather think of them as aberrant. We cannot accept aging as the norm; we want to hide it.

The digital theory of life and death – the theory that life goes from zero to one at birth and from one to zero at death – is historically derived from the Western concept of the soul, according to which the soul enters the body at birth and exits at death. Just as importantly, the concept of soul implies that the self is a single, indivisible entity – rather than a collection of components – and that it continues unchanged (one might say: unsullied) throughout life. Even for those who do not believe in life after death, the Western concept of soul underlies our belief that some essence of ourselves is unchanged throughout life, and that the self is an all-or-nothing kind of thing. However, the concept of soul differs radically from the digital theory of self in asserting that some form of life continues after death, although in what form varies greatly in different cultures.

The Egyptians conceived of a dual soul: the ka survived death but remained near the body, while the spiritual ba proceeded to the region of the dead. The Ancient Greeks had diverse concepts of the soul: some considered it an ethereal substance or abstract property, while for example Epicurus considered the soul to be made up of atoms like the rest of the body. The early Hebrews of the Old Testament refer to the soul as breath, and crucially do not consider the body and soul as separate. Christian concepts of the body–soul dichotomy originated with the ancient Greeks, particularly Plato, who thought the soul was an immaterial and incorporeal substance that animated the body.

It is interesting and revealing that in most ancient civilizations the soul (and life) was identified with breath. The breath/

soul/energy was 'ka' in Eygpt, 'Chi' in China, 'thymos' or 'pneuma' in Greece, 'prana' in India, and 'ruh' in Arabia. The Greek 'pneuma' became the Roman 'spiritus' and thus heir to the Christian 'spirit'. The first entry and final exit of breath from the body were synonymous with life and death. In Greek legend the first man was fashioned by Prometheus from earth and water, but the soul and life were breathed into him by Athena. If breathing is prevented, it leads to loss of conscious-ness and finally death, and thus it would have been obvious that life depended directly on breathing. Words and audible signs of emotion are invisible and can be considered to come from the chest in the breath. In semi-literate cultures thought was often considered to be a kind of talking, perhaps because much thinking and reading was done out loud. And as talking and expressions of emotion were connected with breathing, then thought and emotions could be associated with the breath in the chest.

Breath may have been important to conceptualizing life and soul in another way. Breath is (usually) invisible, yet when we blow hard it can move things and we can feel its pressure on our hand. In this respect it is like the wind, which was often conceived of as the breath and will of gods. Thus, breath was an invisible source of movement *outside* the body, and might therefore act as an invisible source of movement *within* the body, to move the limbs and vital functions.

In the European Middle Ages, St Thomas Aquinas returned to the Platonic concept of the soul as a motivating and ani-mating principle of the body, independent but requiring the substance of the body to make an individual. To René Des-cartes in the seventeenth century, the soul had withdrawn from the body to the brain, and become more-or-less equiva-lent to the mind; hence the body and soul were distinct sub-stances, but could act on each other. To William James at the beginning of the twentieth century, the soul as such did not exist at all, but was merely a collection of mental phenomena.

The first attempts to localize the soul go back to classical antiquity. The soul had originally been thought to reside in the liver, an organ to which no other function could, at that time, be attributed. Homer appears to locate the soul in the lungs/ chest area. Empedocles, Democritus, Aristotle, the Stoics and the Epicureans had later held its abode to be the heart. Other Greeks (Pythagoras, Plato and Galen), together with more modern thinkers, opted for the brain. Descartes famously narrowed down the soul's residence to a small gland within the brain: the pineal gland.

Just as there have been different concepts of the relation of the soul to the body, there have been diverse ideas about when the soul comes into existence and whether it dies. Pythagoras apparently held that the soul was of divine origin and existed before and after death. Plato and Socrates also accepted the immortality of the soul, while Aristotle considered this true of only part of the soul, the intellect. Epicurus believed that both body and soul ended at death. The early Christian philosophers adopted a Platonic concept of the soul's immortality, but thought of the soul as being created by God and infused into the body at conception or birth. In some Christian theology the soul is that part of the individual which is of divine origin, borrowed or created at birth and taken back at death – the divine soul-stuff remains immortal. However, this idea of merging with the divine at death is more of an Eastern thing, and in practice Christianity (and Christ) promised survival after death (or at the time of the second coming) of the individual as an individual, and even the resurrection of the body.

In mediæval Christianity, a soul (separate from the body) was not needed, because the dead were dead until they were resurrected in bodily form, so there was no need for a wispy soul floating around without its body. But the theological innovation of Purgatory, and the scientific exploration of the body, increasingly required theologians to come up with a soul in less intimate contact with the body. In post-mediæval Christian thought, this soul increasingly became the real essence of the individual, and the body became something of secondary importance that could be discarded. By the nineteenth century, many considered the body and soul as separate even during life.

The soul leads to the idea that the self is digital: all-or-nothing, unchanging, and a unity without parts. Opposed to this we have the idea that the self is nothing but its parts, and changes throughout life in both quantity and qualities. A hybrid scheme would have the self (or modern soul) consisting of an essential, digital core that does not change, surrounded by a periphery of the non-essential self. Does it matter which scheme we believe in? Yes – absolutely – it matters to everything human.

If the self is unified and unchanging, then there is no point in adding to it, and there is no consequence of subtracting from it. If the self is completely malleable, then there is everything to be gained by adding to it, and everything to be lost by adding the wrong things or subtracting from it. Here is one of the origins of the differences between the political right and left in politics. The left believes in change, progress, education, society and the primacy of nurture over nature. The right believes in conservation, stability, the individual and the primacy of nature over nurture. If the self is unchanging, society cannot add or subtract from the individual, but if the self is fully malleable and has no core, then society can not only construct it (for good or bad) but also forms part of it. On the other hand, the left believes that all selves are equal and essentially the same, whereas the right believes in the existence of differences between people. So the digital/analogue view of self cuts across the left/right divide in politics.

The consequences of the digital and analogue concepts of self go much further than politics, as they inform our core aims and motivations in life as individuals. Should we seek to change ourselves? Does it matter what we do, watch, read or hear? Should we contribute to society? If the self is not digital (all or nothing), this implies that we can lose parts of the self, for example by forgetting or by changing part of the body or mind. It also suggests that parts of the self may be given to or taken from other people. Whichever way you look at it, it matters whether the self/soul is digital or not.

6
digital self

The concept of self and the concept of death are inextricably linked, because we think of death as the loss of self. If an analogue theory of life and death is replacing a digital theory, then we will need an analogue theory of self too: a self that develops through childhood, but disintegrates with aging or dementia. But this seems paradoxical, because we have been brought up to believe that the core self does not change with age, and we can't have half a self.

The history of the self is hard to trace, but it is thought that following the Classical period of Greece and Rome, during the Middle Ages in the West, the importance of the individual self was downgraded relative to the importance of God and society. The individual had no right to assert himself against God and society; his individual needs were unimportant or even evil; he was there to serve the needs of God and society. During the Renaissance the self seems to have started to reassert itself. The centre of gravity descended from heaven to earth, with man and his nature becoming the primary focus of attention. In his influential *Oration on the Dignity of Man*, written in 1486, Giovanni Pico expresses a view of man that breaks radically with Greek and Christian tradition: what distinguishes man from the rest of creation is that he has been created without form and with the ability to make of himself what he will. In this way man's distinctive characteristic becomes his freedom; he is free to make himself in the image of God or in the image of beasts.

During the Renaissance the individual self burst out from a thousand years of suppression in a multitude of forms. The names of individuals appeared on paintings and books, revolting against the previous anonymity. The ordinary lives of individuals became a suitable subject for books, paintings, laws and academic study. And ordinary deaths of individuals were first marked with gravestones, and their fates after death became subject to intercession by the living. It seems incredible now that the Middle Ages did not think the individual worthy of such attention. The Reformation saw further emphasis on the individual self: each individual, rather than the Church or State, was to be responsible for his own fate, and the individual's experience was more important than the collective deeds or structures in society.

The Enlightenment asserted the supremacy of the individual in a context of equality, Thomas Hobbes' Social Contract suggesting that the origin of all political power should be the people's will, Jeremy Bentham asserting that the ultimate value was the maximum benefit for the most individuals, and Thomas Paine's *The Rights of Man* putting the case for individual human rights. The rise of capitalism meant everyone pursuing their own economic interests as best they could, being responsible for their own income and expenditure, as opposed to the mediæval manorial and abbey system, where work, play and consumption were decided from top-down rather than bottom-up. Adam Smith in *The Wealth of Nations* proposed that society's goals were best served by individuals pursuing their own economic interests. Romanticism asserted the primacy of individual experience in opposition to the mass of society, but was more interested in exceptional individuals and experience rather than equality.

The twentieth century saw two challenges to Individualism. Fascism rejected the spirits of the American and French Revolutions with their emphasis on individual liberty and the equality of men and races, and asserted the individual should be subordinated to the state and its leader, and that individuals, classes and races were intrinsically not equal. Communism, in contrast, asserted that all individuals were equal, but that the interests of society should have priority over those of its individuals, and society rather than individuals should make the decisions. Ultimately both Fascism and Communism were defeated, and Liberal Capitalist Democracy won out. However, in the twenty-first century religion has re-emerged as a competing faith to individualism, or individual selfishness as some might characterize it.

The future may see the disintegration of self. James Hughes, in the *Future of Death*, has prophesied:

> Threats to the self will develop in many areas. Our control over the brain will slowly make clear that cognition, memory and personal identity are actually many processes that can be disaggregated. We will have increasing control over our own personalities and memories, and those of our children. Full nano-replication of the mental process opens the possibility of identity cloning, distributing one's identity over multiple platforms, sharing of mental components with others, and the merging of several individuals into one identity.... When one can easily modify, borrow, or drop, merge with others, and separate, any of their external or internal features, there won't be distinct lines between individuals anymore.

The predominant modern concept of self in the West is of a unified, monolithic entity, changing little with age and differing little with gender, race and the individual. I will characterize this as the atomic view of self, which is the equivalent of the digital theory of life. The digital theory of life indicates that you are either alive or dead, with no degrees of life, and therefore that everyone has more-or-less the same level or type of life. The atomic view of self sees the self as a single, unified entity rather than a conglomeration of different processes, and therefore that it differs little between stages of life and individuals. In addition, the atomic view sees each self as completely separate, i.e. there are no shared selves or shared elements of self. The atomic view of self and the digital theory of life are obviously related and dependent on each other. But is it true that the self is atomic?

At the beginning of the twentieth century, just when the atomic theory of matter was beginning to crack, Sigmund Freud split the atomic theory of self by showing that it had different parts, and that we do not have immediate access to all the different bits. Indeed much, perhaps most, of ourselves lies hidden deep in the unconscious. Subsequent psychological, neurological and neuroscience research has confirmed that the self can be analyzed into different processes subsequently located in different parts of the brain, and that nowhere within the brain is there a central controller or experiencer of all these processes.

In an article on the future of death, James Hughes has written:

> Despite our every instinct to the contrary, there is one thing that consciousness is not: some entity inside the brain that corresponds to the 'self', some kernel of awareness that runs the show, as the 'man behind the curtain' manipulated the illusion in the Wizard of Oz. After more than a century of looking for it, brain researchers have long since concluded that there is no conceivable place for such a self to be located in the physical brain, and that it simply doesn't exist.

Neurologist Oliver Sacks described in his book *The Man Who Mistook His Wife for a Hat* how the loss of particular parts of the brain leads to the loss of particular parts of the self. The inevitable conclusion is that the self consists of parts, and therefore we can have part selves. As evolutionary psychologists Steven Pinker puts it in *How the Mind Works*:

> Our minds are not animated by some godly vapour or single wonder principle. The mind, like the Apollo spacecraft, is designed to solve many engineering principles,

and thus is packed with high-tech systems each contrived to overcome its own obstacles.

What then is the self? What are we referring to when we talk about the self, ourselves or other selves? We may be referring to three different types of thing. The first is the stream of consciousness that we experience from moment to moment: the stream of thoughts and feelings, sights and sounds, images and memories that can only be experienced from the first person perspective. We can think of the experiencer of these experiences as the self. However, it is not necessary for these experiences to have an experiencer. Of course, subjective experiences must be located in a mind attached to someone's body, but there does not have to be a self somewhere within the brain that serially experiences the experiences. Indeed, as many philosophers have pointed, out the idea of an experiencer within the brain experiencing experiences leads to an infinite regress, as how is this internal experiencer to experience experiences without in turn having an experiencer within the internal experiencer!

Experiences do not require internal selves to experience them; rather, they exist independently within minds, and we can think of the self as the sum total of those experiences. Thus the self can be thought of as constituted of experiences, rather than the other way around. To borrow an analogy from Margaret Thatcher: the existence of individuals is not dependent on society, but rather the other way around: society is nothing over and above the sum of its individuals. Analogously, the existence of experiences is not dependent on the self, but rather the other way around: the self is nothing over and above the sum of its experiences, thoughts, feelings etc.

However, something is missing from this collectivist account of self. I do not feel like a collective entity, I fell like a unified self. I think and feel and make decisions from a single perspective: mine. If I listen to what goes on inside me I do not hear multiple selves arguing with each other from multiple perspectives. Rather I hear a single monologue trying to understand all the experience that is presented and integrating everything into a single perspective, which is used to initiate action. This is the stream of thought reported for example by Virginia Woolf's Mrs Dalloway; and referred to as the 'internal monologue'.

Modern psychology recognizes the internal monologue as central to the construction of self. The internal monologue is you thinking to yourself. It is prattling on almost all the time: commenting on experience, commenting on the neighbours, daydreaming, calling up memories, saying how great/rubbish you are, deciding to act, deciding not to act. It is a module of mind that operates mainly in language, and may therefore be located in or in association with language areas of the brain. The internal monologue is the vehicle of self-consciousness in the mind, while the whole range of subjective experience, including sensory experiences and memories, is consciousness.

Self-consciousness is equated with shyness in normal discourse, but in psychology/philosophy it means awareness of or thoughts about internal mental experience rather than external, physical things. However, it may be that shy people have a more active self-consciousness/internal monologue, while extrovert types are more focused on the external world.

In Julian Jaynes's book *The Origin of Consciousness in the Breakdown of the Bicameral Mind* (1976) he made the startling suggestion that self-consciousness first arose only about 3,000 years ago, when Greeks acquired the ability to internalize language, i.e. think in language form. Thus he proposed that some individuals are not self-conscious even now, for example all babies, since they don't have language. The implication would also be that different cultures and different individuals would have different levels (or different types) of self-consciousness, because their different language experience and experience of internalizing language in thought would result in the development of different internal monologues. For example, your self may be completely different from my self because our different experience of language results in different types of internal monologue.

Schizophrenia has been suggested to be primarily a pathology of the internal monologue: either because the internal monologue has become too intrusive and 'loud' or because schizophrenics attribute internal thoughts as external in origin and thus hear 'voices' which may command them to do things. Jaynes suggested that such psychology was more common in the past, and might have accounted for the belief in gods speaking directly to people.

The internal monologue thinks about everything and everybody, including the person within whom the internal monologue is located, i.e. the self. It thinks some people are brilliant and other people are crap, and it may judge the self similarly. Of course it does not think of itself as separate from the person it resides within (at least not in sane people). The internal monologue is continually assessing how well the person is performing, what the person can do well and what not so well, and how well the person performs relative to other people. These judgements form part of the individual's concept of self. But this self-concept of self contains many other miscellaneous entries.

When we think of a particular other person we know, say our mother, we call up many images, memories, beliefs, feelings and judgements associated with that person: the totality of these associations is our concept of that person. Similarly, when we think of ourselves, we may evoke a kaleidoscope of feelings, memories and judgements that constitute our concept of ourselves. The internal monologue is an important part of the experiential self, but it also helps to construct the constructed self.

The self can refer to a variety of different things, but one of them is the experiencer of my experiences, the thinker of my thoughts, the feeler of my feelings, the author of my internal monologue. However, according to current understanding there is no such experiencer of my experiences etc. No such experiencer can be found in the brain – there are just experiences that trigger other experiences, thoughts that trigger other thoughts. But experiences are intrinsically subjective – they are experienced from inside as mine – I can't experience someone else's experiences without becoming that someone else. According to this scheme of things, the self is an illusion, another convenient fiction, like society or the average voter.

This concept of self as the illusory subject of experience has been worked out by many psychologists and philosophers. And the concept is also at the core of Buddhist ideas of the world. The Buddha considered the self, soul and even external reality an illusion. Instead, human existence is a composite of five constituents (body, sensations, ideas, dispositions and consciousness), each of which is an aggregate of other components, and none of which is the self or soul. A person is in a process of continuous change, with no fixed underlying entity. Life is a stream of becoming, a series of manifestations and extinctions. The concept of the individual ego is a popular delusion; the objects with which people identify themselves – fortune, social position, family, body, and even mind – are not their true selves. Nothing is permanent. However, Buddhists accepted the Hindu idea of Karma and rebirth, which, as many non-Buddhist Indian philosophers have pointed out, is hard to reconcile with the concept of no permanent self or soul.

However, we are still missing an important aspect of what we mean by the self. This aspect was pointed out by the Behaviourists, a revolutionary movement of academic psychology starting in the 1920s. The Behaviourists didn't believe in all this wishy-washy navel gazing of introspective psychology. Your behavioural self is your tendency to behave in certain ways. This includes your tendency to do particular things, say particular things, emote or vote in particular ways, and express particular beliefs in particular situations.

When we think about what characterizes someone else's self (rather than ourselves), it is these kind of *behavioural* things we think of (rather than their internal experiences or internal monologue). This behavioural self includes their 'character' and 'personality', although these terms refer specifically to the behaviour that distinguishes the person from other people, rather than the total behavioural self. When we think about other people's selves we think about the behavioural self because that is all we have access to, whereas when we think about our own self we can include both our constructed self, which includes how we view our own behaviour, and our experiential self, which we alone have access to.

Behaviourism faded away in the 1950s and 1960s, when the success of the computer and neuroscience led to renewed interest in the brain and computational theories of mind. However, towards the end of the twentieth century, continental philosophy and sociology criticised the Western dualistic concept of self, i.e. that there were separate material and non-material aspects of self or mind. This interestingly led to a re-emphasis on the body (rather than the brain or mind) as the site of the self, so that changes in the body, due for example to fashion, disease or aging, automatically corresponded to changes in the self. And this harps back to William James's theory that emotions are located in the body, not the brain. So laughing and crying, fear and anger, are produced and experienced primarily in the body, not the mind.

Thus there are three rather different aspects of self. The experiential self is the stream of consciousness that constitutes my present subjective experience, while the constructed self is my concept of myself as a person in relation to other people. And the behavioural self is my tendency to behave in certain ways.

These ideas of self are very different from that of a mono-lithic, immutable soul or spirit, the subject of all experience and actions. However, it is conceivable that the experiential, constructed and behavioural selves are immutable, unchang-ing throughout life (and death). So are they? Our own mem-ory suggests that our moment-to-moment experience of life was different when we were babies, when we were children and now. And we know that people's psychological capacities, for memory, reaction, and complex thinking are different at different ages, making it unlikely that they would have the same experience. As for the constructed self, it is clear that people's concepts of themselves change with age and experi-ence, although certain aspects may remain constant. Simi-larly, many aspects of our behaviour change with age, although we may be able to identify a limited number of behavioural traits common to the infant and old-aged forms of the same person.

So the self is not immutable: it changes with age. We are more like a wave than an atomic particle. According to classi-cal ideas, an atomic particle is immutable, indivisible and the same everywhere it goes, just like the old concept of self. A wave is changeable, divisible and continuously changing. Like a wave we move through life, picking up new experiences, memories and beliefs, but discarding old ones, so that after a time nothing is the same within the wave except the form of the wave, and that too may change.

The body and brain may seem like an anchor maintaining the identity of the self, but as Steve Grand points out, this may be an illusion. In his book *Creation: Life and How to Make It'* he asks us to think of an experience from memory:

Something you remember clearly, something you can see, feel, maybe even smell, as if you were really there. After all, you really were there at the time, weren't you? How else would you remember it? But here is the bombshell: you *weren't* there. Not a single atom that is in your body today was there when that event took place.... Matter flows from place to place and momentarily comes together to be you. Whatever you are, therefore, you are not the stuff of which you are made. If that doesn't make the hair stand up on the back of your neck, read it again until it does, because it is important[1].

Of course, Grand is getting a little carried away here (remember the old schoolboy gobsmacker: every breath we take contains a molecule from Caesar's last breath), but the important point is that the apparent stability of self is not based on stability of the matter making up ourselves. Rather it is based on the relative stability of processes controlling the arrangement of matter within the brain. And these processes change with time. Every time we think or feel something or talk to someone, our self and its brain changes a little bit.

Every morning we wake up and reboot the old self, but each morning the old self is a little bit different from what it was before. So that over time we change – like a wave. If we remembered everything that had happened to us in the past, and memory was a direct link to the past, then the present self would be directly anchored to the past self. But we do not remember everything, and memory is not a direct record of the past; rather it is a present reconstruction from decaying fragments of the past.

We generally vastly overestimate how much of our lives we can remember. Pick a random date from your distant past, and try to remember something that happened on that day. The chances are that you will remember nothing. In fact, the chances are that you will only remember a few 'events' from that year, and those events will be in a narrativized form lacking any detail of the experiences themselves. Just as visual images and videos take up a lot of computer memory, they also take up a lot of space in brain memory, and so events tend to be reduced to a narrative sketch with a few thumbnail images.

Memory scientists have realized the central importance of forgetting, because the total memory capacity of a human is massively dwarfed by the total experience capacity. We remember almost nothing of our lives. Each night we are thought to have on average five dreams – five voyages of the imagination – but how many are remembered? Five dreams every night of your life adds up to at least 50,000 lost dreams. Add to that the rest of your forgotten daytime life. This is almost a holocaust of experience. And only certain types of things can be remembered. If we could really remember pain we would be in pain again. If we could really remember falling in love, we would be falling again. If we could really remember ecstasy, why would we do anything else? Memory is not a recall of the thing, but rather a registering that the thing happened. Memory is a very tenuous connection to the past. Because it is so tenuous the present self is partly independent of the past self.

Even though memory is tenuous, it is vital. Without it there could be no constructed self at all, only an experiential self, existing only in the present, cut free from all connection to previous or future selves. This is the haunting image from the film described previously: the brain-damaged spastics living entirely in the present, like children, like Alzheimer's patients. Loss of memory and the ability to form new memories is one of the most debilitating aspects of aging. It slowly deprives us of names, of language, of a context in which to talk to people, of meaning, and finally of reason.

Memory binds things together, and without it, things fall apart. The constructed self rapidly crumbles without memory, and the experiential self loses its coherence and starts to drift apart. The internal monologue loses its context for commenting on things without memory; indeed, it could conceivably be lost altogether. However, memory loss is not all doom and gloom. People who have completely lost their memory due to brain damage generally appear fairly happy. Loss of memory can be a release for some people, allowing them to live in the present. In some ways complete memory loss approximates to Nirvana. But not the Nirvana we were hoping for.

We may think that very old people, and very young people, and people from very different cultures are different on the outside, in their appearance and capacities, but are basically the same on the inside, in how life is subjectively for them. Unfortunately we cannot know what it is like to be somebody else, any more than we can know what it is like to be a bat. The only way to communicate what it is like is language, and language is notoriously bad at communicating such things. To paraphrase Wittgenstein: Of those things which one can not speak, one must pass over in silence. But one can guesstimate some aspects of internal experience from psychological capacities, from memory, reasoning, reaction times, and language. These and other investigations suggest that young children may live in a very different world from us. Psychological investigations of old and very old people indicate that they are very heterogeneous. No two are entirely alike: some appear little different from middle-aged people, while others appear to live in a different world. It is impossible to know what it is like to experience Alzheimer's disease from the inside, or another dementia, or extreme old age – but it is worth thinking about because that is where we are all headed.

Some may think that the self stays the same with age, but I would contend that the self is not one thing and that its contents change with age; only the form remains roughly the same – like a wave that rises up, travels the oceans and then fades away, or alternatively crashes down on the shore, only to feed into other waves. A wave may have roughly the same form as it moves across the ocean, but the molecules of water that make it up are continuously changing. Analogously, we appear to have roughly the same form from the outside from year to year, but the molecules of experience that constitute our internal life are continually changing, and our internal selves are slowly aging just as the body is aging. Our form is not immortal or even constant, but like a ripple in a pond it may affect other forms that outlive us. Ultimately we are all one big pond of experience; it's just that while we are alive we identify with only one ripple.

The concept of shared selves may seem like an oxymoron. But if we consider, in this globalized, homogenized world, how much of our experience, upbringing, education, television and media exposure is common (despite the differences), it is hard to see how we can avoid ending up with common elements to ourselves. This is the basis of our common culture. Any two people who have shared the same education, the same TV programme, the same idea, or the same deeply emotional or religious experience, will have similar elements of self stored within them. Some people, such as family members or twins, will obviously end up with more common elements to their selves than others. If people share common elements to their selves, and we reject a unitary concept of self, then inevitably we need to consider the possibility of partially shared selves.

In 1965, Robert Ettinger wrote the manifesto of the cryonics movement, *The Prospect of Immortality*: 'The simplest conclusion is that there is really is no such thing as individuality in any profound sense.... Let us then cut the Gordian knot by recognizing that identity, like morality, is man-made and relative, rather than natural and absolute'.

Identity is crucial here. I am what I identify with. Some schizophrenics may apparently think of some of their own thoughts as belonging to other people. Or they may believe that more than one self occupies the body. People with spinal cord injuries can lose identification with their limbs. Surgical cutting of the corpus callosum that connects the two sides of the brain apparently results in two selves each experiencing and controlling one side of the body only. When I get up in the morning I identify with the person who went to bed last night, because I have a continuity of memory – if I didn't remember being that person I wouldn't identify with them as myself. But that person last night doesn't have to identify with me, the person he may be in the morning – that's probably why he got blind drunk and left me with a hangover. Come to think of it I'm not entirely sure what he got up to last night!

And what about the person who was the self of the five dreams I had last night? Should I identify with him/her, or the five dreamers, or the 50,000 lost dreamers of my life? Should I identify with my children or ancestors, my ideas, my football team, my local community, or my community of fellow believers? The point is that there is no correct answer to these questions. Who I identify with defines the boundary of the self, not the other way around. My identity distinguishes between self and non-self, and it distinguishes what I should care for or protect (everything inside the loop) from 'the other' (everything outside the loop). How I regard my own death depends, in part, on what I identify with.

As Richard Dawkins points out in *The God Delusion*, there is nothing special about the old man dying – he, and we, have done it several times before. The child that was him, 'died' long ago, by growing into an adult. Shakespeare's seven ages of man each 'die' in turn by morphing into the next phase. The Christian who 'dies' and is 'reborn' does so by losing their identification with their previous self, and re-identifying with the life of Christ. The car-crash victim who loses her memory is also 'reborn'. In each of these cases we may be tempted to think that something has stayed the same through the change, and therefore the self has remained intact. But this is 'digital/atomic' thinking: we are thinking that self cannot both change and stay the same.

If we accept that the self is made of components, then obviously some components can change while others stay the same through life. We may choose to think that those components that stay the same are the 'essential' self, but if so this choice is rather arbitrary. Of course, it is your choice what you identify with. But the old man may choose to identify with the wisdom gathered over a lifetime that distinguishes him from the child and youth he was. And the born-again Christian chooses to identify with the new self and new ideas, not the old. A self made of components can also identify with the components, irrespective of which self the components occupy (or constitute). So we may identify more with our children, our homes, our country or our ideas rather than our selves. And obviously these things may survive the death of the individual self.

What we are discussing here may seem abstruse and abstract, but it impacts directly on the digital theory of life and death. According to the atomic theory of self, the core self is unchanged throughout life, and either vanishes to zero at death, or continues unchanged as the soul. But according to the wave theory, the self is a looser collection, whose composition changes throughout life and after.

A digital theory of death leads to a digital theory of life: you are either fully alive or fully dead, with no shades, varieties or different ways of being alive or dead. And a digital theory of life leads to a digital theory of self: a self either exists or it does not – there are no degrees or components of self – all selves are equivalent and unchanging, at least in terms of being selves. But an analogue theory of death leads to an analogue theory of life: there are different degrees and ways of being alive and dead. And an analogue theory of life leads to an analogue theory of self: selves have degrees, components and varieties; they rise up, swap components, change from day to day, and finally fade away.

interlude 6
the fear of death

To be, or not to be: that is the question:
Whether 'tis nobler in the mind to suffer
The slings and arrows of outrageous fortune,
Or to take arms against a sea of troubles,
And by opposing end them? To die: to sleep;
No more; and by a sleep to say we end
The heart-ache and the thousand natural shocks
That flesh is heir to, 'tis a consummation
Devoutly to be wish'd. To die, to sleep;
To sleep: perchance to dream: ay, there's the rub;
For in that sleep of death what dreams may come
When we have shuffled off this mortal coil,
Must give us pause: there's the respect
That makes calamity of so long life;
For who would bear the whips and scorns of time,
The oppressor's wrong, the proud man's contumely,
The pangs of despised love, the law's delay,
The insolence of office and the spurns
That patient merit of the unworthy takes,
When he himself might his quietus make
With a bare bodkin? who would fardels bear,
To grunt and sweat under a weary life,
But that the dread of something after death,
The undiscover'd country from whose bourn
No traveller returns, puzzles the will
And makes us rather bear those ills we have
Than fly to others that we know not of?
Thus conscience does make cowards of us all;
And thus the native hue of resolution
Is sicklied o'er with the pale cast of thought,
And enterprises of great pith and moment
With this regard their currents turn awry,
And lose the name of action.

William Shakespeare, *Hamlet*

Hamlet is torn between action and inaction. If he chooses inaction, he must accept an outrageous fortune: his father's murderer lying in his mother's bed, and usurping his throne. If he chooses action, to kill the king or himself, he must face death. But death is an undiscovered country that may be nothing (sleep) or may be something far worse that life (a nightmare). Shakespeare suggests that the fear of death prevents an honourable life and makes cowards of us all.

The dread of death and non-existence appears to be universal and is recorded throughout history. Evolutionary psychologists suggest that it is a primordial fear built into our genes to promote survival at all costs. Recipes for overcoming that dread are multiple. In the Gilgamesh epic and many others it is suggested that leaving a great or good name is what matters. A number of ancient and modern commentators have suggested that the solution is sex and babies. Religions offer a get-out clause by revealing a hidden life after death. The Greek philosopher Epicurus considered the fear of death as one of the greatest sources of unhappiness in life. But he thought the fear of death groundless as 'death is nothing to us, so long as we exist, death is not with us; but when death comes, then we do not exist'. Mark Twain dismissed the fear of death thus: 'I do not fear death. I have been dead for billions and billions of years before I was born, and had not suffered the slightest inconvenience from it'.

If death is a permanent loss of consciousness, there is no point worrying about death, because when dead we won't be aware of it. In former times it was thought possible to be conscious when dead, so it certainly was worth worrying about. But in the absence of consciousness there is no continuity of self between the alive and dead body, and therefore no more point in worrying about being dead than worrying about being someone else or a dog or a rock. So according to this line of reasoning you never will be dead because being dead is not a state that you, yourself, can be in. As the Austrian/Cambridge philosopher Wittgenstein wrote in the *Tractatus Logico-Philosophicus*: 'Death is not an event in life. We do not live to experience death. Our life has no end in just the way in which our visual field has no limits'.

However, as the Greek comic Epicharmus quipped: 'I am not afraid of being dead, I just do not want to die'. People are not just afraid of what might come after death, but also and perhaps more potently of non-existence itself. Against this, Epicurus asserted that life is lived in the moment, and pleasure is not increased by contemplating its continuation in the future; just the opposite, in fact: pleasure would pale in eternity. Essentially Epicurus, and many others following him, are suggesting: forget death, live for the moment. More recently, the philosopher Thomas Nagel has argued against Epicurus: 'If we are to make sense of the view that to die is bad, it must be on the grounds that life is good, and death is the corresponding deprivation or loss, bad not because of any positive features but because of the desirability of what it removes'. Thus according to Nagel, it makes sense to fear death because death removes what we have and want, life.

The poet Philip Larkin expressed the same thing in his poem 'Aubade' thus:

This is a special way of being afraid
No trick dispels. Religion used to try,
That vast moth-eaten musical brocade
Created to pretend we never die,
And specious stuff that says *No rational being
Can fear a thing it will not feel*, not seeing
That this is what we fear – no sight, no sound,
No touch or taste or smell, nothing to think with,
Nothing to love or link with,
The anaesthetic from which none come round.

How does the failure of the digital theories of life and self impact on the fear of death? In many ways it makes things worse, because death gets drawn into life with aging, so that we can no longer dismiss death as a state we will not experience. Death is no longer out there waiting at the end of the tunnel; it is inside all of us all the time. And death will grow inside us until it bursts us apart. But the news is not all bad. Death within life can seem tamer, more natural, less drastic. Death occurs every time we go to sleep (the 'little death' of Gilgamesh), as dreamless sleep is non-existence. And if every day we wake up a slightly different person, so that the self is not continuous through life, it may not matter so much if one day we (or the somebody else we have become) do not wake up at all.

Moreover, if the self is not a single, unified entity, but a wave constituted of genes and memes, then the dissolution of the self at death does not necessarily mean the end, because the genes and memes may continue in other selves. The end of binary life/death does not just mean that death gets mixed up in life, but also that life gets mixed up in death. And immortality becomes possible, as we shall see.

7
cellular death

Our body is a community of trillions of individual cells. The lives and deaths of that community and its individuals are distinct but intimately connected. Death of individual cells may cause death of the body, and death of the body eventually kills all the individual cells. But cell death also occurs naturally all the time without damaging the body; indeed, certain types of cell death are essential to the development and survival of the body. The mature brain is sculpted by the programmed elimination of redundant cells, the womb lining is eliminated partly by programmed death every month, and virally infected cells commit suicide to prevent the spread of infection. So death is essential to life.

Almost every cell in the body has a primed bomb within it, a programme of self-destruction, ready to be triggered by any deviation from the norm. And almost every cell has a timer within it that counts the number of cell divisions, and after a fixed number stops dividing and initiates a programme of senescence. Self-destruction is the default mode of many cells, unless they get continuous reassurance from other cells around them that they are doing the right thing.

However, in rare cases, individual cells may escape the mortal fate of the body. Germ cells, sperm and eggs, can escape their parent body to form new selves and new germ cells. Cancer cells are potentially immortal, as they have inactivated the normal programme of senescence and self-destruction. In my laboratory we grow cancer cells originating from a black American named Henrietta Lacks, who died from cancer in 1951, but her cancer cells (known as HeLa cells) are grown in biological research laboratories all over the world, forming a collective mass hundreds of times that of their original owner. Her cells have been into space, used to develop the polio vaccine and have advanced many branches of science. In recognition of her contribution to science the city of Atlanta has named 11 October Henrietta Lacks Day. Her cells and her name are immortal!

The idea that cells may commit suicide goes back to a scientific paper published in 1972 by John Kerr, Andrew Wyllie and Alastair Currie, which caused little stir at the time, but has since achieved mythical status in the scientific community. Kerr, Wyllie and Currie were pathologists simply looking at the shape and form of cells from diseased tissues of the body. This, of course, had been done by countless pathologists for at least a hundred years. But the painstaking analysis of Kerr, Wyllie and Currie showed them something that no one else had noticed, or at least not thought important: cells in diseased tissues appeared to die in two entirely different ways. Some cells appeared to explode: an explosion starting inside the unfortunate cell, which became chaotic and swollen, and then spreading to the outside as the cell swelled until it burst its cell membrane (the thin wall encompassing the cell), spilling the contents of the cell out into the surrounding tissue. This wreckage of the exploded cell was damaging to neighbouring cells, causing them to become inflamed.

Inflammation is a normal response to tissue damage, recruiting immune cells and processes to sterilize the damaged area of invading micro-organisms and helping to repair tissue. But inflammation is a radical and expensive response to tissue damage, which because of the toxic agents used to destroy pathogens may also damage our cells. This explosive form of cell death is called 'necrosis', and because it damages surrounding cells and induces inflammation, it is in general undesirable and to be avoided if at all possible.

Kerr, Wyllie and Currie also noticed a less obvious and almost silent form of cell death that others had overlooked, which they called 'apoptosis' (Greek for 'falling off', as leaves in winter). The cell did not swell, but rather shrank, and divided itself into bite-sized pieces that were promptly eaten by passing phagocytic cells, which left no trace of the cell's demise behind. Phagocytes (Greek for 'eating cells') are scavengers that normally eat bacteria and cell debris, but apoptotic cells put out signals on their surface asking phago- cytic cells to eat them. It was partly because of the rapid eating of these apoptotic cells that this form of cell death had not been noticed before.

Apoptosis was a benign form of cell death that did not damage surrounding cells or cause inflammation, and throughout the silent process the contents of the cell re- mained organized and functional, participating actively in its own demise. Apoptosis is a form of cellular suicide, and it was a very big hit in the scientific community, making the formerly staid subjects of pathology, toxicology and cell death sud- denly sexy.

Subsequent to Kerr, Wyllie and Currie's ground-breaking study, the cellular mechanisms underlying apoptosis were elu- cidated by many different scientists using many different approaches. Studies of apoptosis in the nematode worm by Sulston, Brenner and Horvitz in the 1980s and 90s gained them the Nobel Prize in 2002. Sydney Brenner was one of the grand old men of molecular biology: he had helped unravel the genetic code and how it was translated in cells back in the 1960s. Having worked out how genes control the cell, he became interested in how genes determine the structure of the multicellular organism. How does the genetic programme dictate how a single egg cell will develop into a hugely com- plex multicellular adult? To tackle this question, in the 1980s he turned to something relatively simple: a tiny transparent nematode worm.

Nematodes, more commonly known as roundworms, are a class of animals of which there are at least 15,000 currently identified species, but probably hundreds of thousands of unidentified species. They are one of the most abundant ani- mals in the world, occurring in most habitats, including inside the bodies of animals and plants as parasites. Thankfully most of them are small, although parasitic nematodes of whales grow up to 7 metres long. The species of nematode studied by Sydney Brenner, known affectionately as *C. elegans*, is less than a millimetre long, and the adult consists of exactly 959 cells. How did it develop from one egg cell to 959 adult cells, each slightly different from the others and located at a particular spot within the integrated adult body?

This is where the transparent nature of the nematode and the angelic patience of John Sulston came into the story. He came to work in Brenner's lab in the famous Cambridge Laboratory of Molecular Biology (LMB), which has incubated 15 Nobel Prize winners so far. John Sulston went on to sequence the worm's genome, and subsequently headed the British end of the publicly funded drive to sequence the human genome, for which he was knighted. Back in the 1980s, as a young post-doctoral researcher, Sulston simply looked at individual nematodes developing under the microscope for hours, days and weeks on end. The worm only took 14 hours to develop from an egg to an adult, but following the fate of the cells was like trying to a do a thousand-piece puzzle, where the pieces are more-or-less identical, microscopic, transparent, and rapidly moving in three dimensions. A nightmare then, but a nightmare in which one has to stay awake, for a single break in concentration may mean starting all over again. He watched the egg cell divide into two, and the two divide into four, and he watched each and every one of the subsequent progeny move, divide and develop, until he could trace the developmental history of every cell in the nematode's body.

Sulston noticed something odd during this development: not all the cells that were born by division survived to the adult. A total of 131 particular cells disappeared soon after being born by division, leaving room for other cells. And it was always the same cells in each developing worm. These cells were genetically programmed to die: their genes were ordering them to commit suicide. Because the genetic make-up of the worms was relatively easy to alter, it was possible to identify the genes controlling this 'programmed cell death', and thus the particular machinery defined by those genes that executed the death programme within the cell could be discovered. Or at least that's the theory. In practice it can take decades of frustrating lab work.

To attempt this complex mission, Brenner and Sulston recruited Robert Horvitz to Cambridge from Jim Watson's lab at Harvard, and Horvitz completed the task back in the US. It was Robert Horvitz who first showed that programmed cell death in the nematode worms was executed by the same mechanisms that caused apoptosis in ourselves. Thus programmed cell death and apoptosis are one and the same thing, and they both involve shredding of the cellular machinery by molecular scissors called 'caspases'.

How did Horvitz work this out? He exposed the worms to mutagens (radiation or chemicals that mutate the DNA), and looked through thousands of their progeny for mutant worms that had a changed number of dead cells. Eventually he (or rather one of his laboratory workers) spotted a mutant worm where all of the 131 cells that were programmed to die actually lived and survived as normal cells in the adult. Then the laboratory searched through the mutant worm's DNA to find which gene had mutated. Unfortunately it was a gene which at that time had no known function. But comparison of its DNA sequence to other known genes showed that it was similar to a human gene that coded for a caspase. Caspases are molecular scissors that cut up proteins inside cells. Subsequently they put this worm gene into human cells in a dish, and showed that this single gene could cause human cells to undergo apoptosis. Thus a single gene, and the protein it coded for, could determine whether a cell lived or died.

What is the point of having cells commit suicide during development? There are probably many different reasons. Sometimes it seems cell division is required to change one type of cell into another type. But after this 'differentiation' only one daughter cell is required, so the other is recycled. Sometimes a solid structure formed during development needs to be hollowed out or sculpted into a more mature structure. Sometimes a tissue with a particular function becomes redundant during development; for example, the tadpole's tail is eliminated by apoptosis when the tadpole turns into a frog.

Sometimes apoptosis is used to select out a particular set of cells: during early development thousands of genes are spliced together to generate antibodies that recognize almost any shape. But we don't want antibodies that recognize shapes that are already present in the body – otherwise the antibodies will attach to these shapes and direct the immune system to attack them. So during development, those immune cells that generate these dangerous antibodies are ordered to commit suicide by apoptosis. That leaves alive only those immune cells generating antibodies that do not recognize molecules in the body, so they will not attack the body, but will attack anything else such as foreign bacteria or viruses.

Sometimes apoptosis is used to eliminate the weak links in a Darwinian struggle for survival. During brain development, neurons have to connect up to each other in a fantastically complex pattern of networks. Some neurons compete with each other to connect up with target neurons some distance away – those that succeed receive sustenance in the form of 'growth factors' released by the target neurons that prevent apoptosis in the connecting neurons. By contrast, those neurons that do not find target neurons to connect to eventually wither away by apoptosis. Indeed, this is a general mechanism to ensure that particular cells get to particular locations during development and stay there. A cell that wanders away from its developmentally defined location loses the particular mix of signals that it gets from its neighbouring cells in the form of cell contacts and growth factors. And when it loses these signals it commits suicide by apoptosis, presumably because the wrong cell in the wrong place can cause trouble.

One particular kind of trouble that a rogue cell can cause is cancer. For a cancer cell to cause real trouble it has to leave its original place of birth, voyage through the body and found a new colony of proliferating cancer cells somewhere else. But the majority of cells leaving their site of origin will immediately commit suicide as they lose the signals saying they are in the right place. Cancer cells only manage to bypass this major check on cancer development by mutating in ways that subvert the suicide programme or the signals that regulate it. Mutations themselves can trigger apoptosis if the cell detects them in its DNA, and this is another major way of preventing cancer. However, mutations in genes for the machinery that detects mutations or DNA damage can circumvent even this block on cancer. And the importance of this block is shown by the fact that the majority of human tumours have disabling mutations in the cellular machinery that detects DNA damage and mutations, allowing a range of other mutations to accumulate that eventually allow a cancer to develop.

If cancer cells are recognized as foreign by the immune system (some types of cancer cell are recognized as foreign, but many are not) immune cells may induce apoptosis in the cancer cells by spraying or injecting them with apoptosis-inducing molecules. A similar trick is used to eliminate virally infected cells if they are recognized by the immune system. Alternatively some types of virus are recognized by the infected cell itself and cause it to commit suicide, thus preventing spread of the virus. A case of better dead than red? As with cancer cells, viruses may fight back by inactivating the machinery of apoptosis. Or in the case of the AIDS virus (HIV) it may turn apoptosis to its advantage by triggering apoptosis in the key immune cells that are fighting it. This enables the virus to spread, as well as other diseases normally kept in check by the immune system; hence the name Acquired Immune Deficiency Syndrome.

Some heart and brain diseases may trigger too much apoptosis, rather than too little. This leads to the loss of vital cells, which in the case of neurons and heart cells, can not easily be replaced. However, if cells are damaged by disease, it may be better if the cells die rather than live dysfunctionally due to their damage. And if cells must die, it is better they die by apoptosis rather than by explosive necrosis. A damaged neuron may disrupt signalling in the brain, or a damaged heart cell may disrupt the heartbeat, so these dysfunctional cells may best serve the body by committing suicide. In fact, apoptosis can be triggered by many forms of cell damage, including DNA damage, protein clumping or membrane disruption. The cell must make some kind of judgement after damage as to whether it is capable of repair, and if not then activate the suicide programme.

Not all cell death is intentional, and by far the most important cause of accidental and disease-induced cell death is 'ischaemia'. Ischaemia means not enough blood flow to a tissue to supply its oxygen and energy requirements. There are many different causes of ischaemia – the most dramatic is when the heart stops so that the blood supply to every organ, tissue and cell in the body is suddenly brought to a halt. A heart attack is caused by blockage of the blood supply to the heart, and usually kills cells in the heart even if the heart's owner survives. Heart failure is caused in part by insufficient blood supply to the heart. A stroke is caused by blockage of the blood supply to the brain, and usually causes death of brain cells.

Even a cancer tumour normally damages surrounding tissue and organ function by disrupting the blood supply. Pneumonia and other respiratory diseases disrupt oxygen uptake via the lungs into the body and thus its supply to the body. Ultimately, at the terminal event of death, the heart and lungs stop, causing rapid ischaemia throughout the body. Ischaemia and the consequent lack of oxygen are thus the final common cause of disease-induced death in our tissues and cells. Dr Milton Helpern, who was for twenty years the Chief Medical Examiner of New York City, stated it thus: 'Death may be due to a wide variety of diseases and disorders, but in every case the underlying physiological cause is a breakdown of the body's oxygen cycle'.

Without a blood supply of oxygen and glucose, the cell's stores of oxygen and fuel are rapidly used up, and the cell's energy supply dries up. Without energy all the processes maintaining the cell stop, molecules and structures start to break up, and there is no energy to rebuild or repair anything. Of more immediate concern is that calcium and sodium ions leak into the cell and can no longer be pumped back out of the cell, because that requires energy. The entry of sodium causes the cell to swell, and if the oxygen supply is not restored, eventually the cell swells so much that it bursts and dies by necrosis. Alternatively, the entry of calcium triggers destruction of the cell, which may end by necrosis or apoptosis. Either way, a lack of oxygen can trigger cell death, which may take many hours in skeletal muscle, but half an hour in the heart, and just a few minutes in the case of neurons in the brain. And the death of neurons or heart cells is irreversible because even healthy neurons and heart cells cannot divide to replace their dead neighbours. That is why a heart attack or stroke is so damaging: the lack of oxygen very rapidly kills cells in the heart and/or brain, which cannot be replaced.

Our inability to replace neurons might be bearable if they were a relatively robust cell type, but unfortunately the opposite is the case. Neurons are delicately neurotic and seem to drop dead at the merest hint of stress. If the body is deprived of oxygen, by heart attack, respiratory failure or death, the first cells to cave in are the brain neurons: they lose function after just a few seconds of ischaemia (we faint) and many are irreversibly damaged within a few minutes. By contrast other cells hold out much longer: skeletal muscle cells last for hours or days in an otherwise apparently dead body. The reason for this fragility of neurons is two-fold. Firstly, they depend almost entirely on oxygen for energy production, whereas other cells, such as muscle cells, can also generate energy in the absence of oxygen. Secondly, and more interestingly, the brain has a major design fault. The brain's main signalling molecule – glutamate (which is the main means by which neurons talk to each other) – is also a potent neurotoxin!

Glutamate is a neurotransmitter – a chemical transferring signals across the small gap between adjacent neurons. Neurons can conduct electrical signals along their length very rapidly, but the electrical signal cannot jump from one neuron to the next, across the gap between them, known as the 'synapse'. Instead, when the electrical signal reaches the end of the nerve, glutamate is suddenly released from the nerve terminal into the gap. The glutamate released from the end of one neuron activates 'receptors' on the surface of the next neuron.

Cells have many different types of receptors on their surface to detect different signals from outside the cell. Glutamate stimulates glutamate receptors on the outside surface of the next neuron, causing it to generate a new electrical impulse, and thus continue the journey of the electrical signal within the brain. However, the glutamate released out of the first neuron is immediately pumped back into the cell in order to terminate the signal. This pumping of glutamate back into the cell requires a lot of energy. When the energy runs out, the pump runs backwards, so the glutamate floods out of the cell into the gap. Once in the gap the glutamate stimulates all its normal receptors on the next neuron. But it also stimulates a special glutamate receptor, the so-called NMDA receptor, which is the Achilles heel of the brain.

Mild and transient activation of the NMDA receptor allows low levels of calcium into the neurons that facilitate memory formation. But massive activation by high levels of glutamate causes the neuron to overload with calcium, triggering destruction of the cell. And this is why our neurons are killed during stroke or neurodegenerative disease. It also contributes to so called 'Chinese restaurant syndrome' resulting from over indulgence with soy sauce. Soy sauce is stuffed with monosodium glutamate (MSG), a form of glutamate, used almost ubiquitously as a food additive to give a 'meaty' taste. Too much soy sauce results in high levels of glutamate in the blood. Normally this would not bother the brain, as the brain protects its over-sensitive neurons with a protective layer known as the 'blood–brain barrier'. Substances in the blood cannot normally enter the brain without crossing this barrier, which protects the sensitive neurons. But if levels of glutamate in the blood become too high or the barrier is damaged, some of the glutamate squeezes through the barrier and wreaks havoc in the brain.

Glutamate-induced killing of neurons is called excitotoxicity, because it results from the neurons becoming over-excited. Discovery of this death pathway in the brain led to great expectations during the 1990s that drugs designed to block the NMDA receptor would protect neurons against stroke and neurodegenerative damage. But all this expectation came to nought, partly because blocking the NMDA receptor disturbed normal brain function. Indeed, drugs that blocked the

NMDA receptor caused loss of consciousness – leading to the startling claim that the NMDA receptor was the consciousness switch of the brain – turn it on and you are conscious, turn it off and you are unconscious.

All this made it difficult to use the drugs to protect the brain, as not many people wanted to swap a damaged but conscious brain for an intact but unconscious brain. However, we might accept temporary loss of consciousness in order to gain temporary protection of the brain, for example during stroke, epilepsy or brain surgery. Unfortunately, the NMDA receptor-blocking drugs turned out to have relatively little ability to save neurons; and this led to the recognition that other neuronal receptors could contribute to neuronal death. Scientists turned their attention away from the surface of the cell towards the depths below, or rather inside, the cell in order to look for a final common target within the neuron that was controlling neuronal death. If drugs could be designed to hit this target, then we might have a better chance of saving neurons. Fortunately, the scientific detectives seeking out the causes of apoptotic and necrotic cell death had already followed a trail of clues leading back to an apparently unlikely suspect as the central mastermind of cell death – the mitochondrion.

The mitochondrion is a specialized microscopic structure within the cell, of which there may be several hundred per cell. Mitochondria (pronounced 'my-toe-con-dria') can be thought of as endosymbionts: semi-independent life forms living inside our cells for mutual advantage. Our cells provide them with food and shelter, and in return they generate almost all our energy – in the form of electricity. They are thought to have arisen two billion years ago when a bacterial ancestor of mitochondria invaded the ancestor of our cells to form the first modern cell. However, modern cells and their mitochondria have evolved so intimately together that they cannot survive without each other. But mitochondria still retain their own DNA, separate from the main store of cellular DNA in the nucleus, and reproduce independently of the rest of the cell. They look like microscopic worms, and they wriggle around the cell, occasionally fusing with or budding off from other mitochondria. If the mitochondria are specifically stained – so that we can visualize them down a microscope – the cell really does look like a can of worms. Only these worms are alive, wriggling around the cell, and bursting with electricity.

Mitochondria play a central role in initiating both apoptosis and necrosis. They pronounce the death sentence of the cell. This is probably no coincidence, since the mitochondria are the main target for many noxious stimuli, such as toxins, free radicals, excessive calcium and lack of oxygen. The mitochondria are damaged by these fatal stimuli, and the consequence of this damage is that the energy supply to the rest of the cell fails. The knock-on effect is that without energy the whole cell spins out of control. The stupendous juggling act that is cell life depends on a continuous supply of energy from the mitochondria. Without it the cell literally falls apart. The molecules that are meant to be kept outside the cell cannot be prevented from coming in, and the molecules that are meant to be inside cannot be stopped from leaking out.

Since the mitochondria are sensitive to the lethal events causing necrosis, it makes good sense for the mitochondria to be triggers for apoptosis too. That's because if a cell is exposed to a lethal event (such as a toxic dose of radiation), it would be best for the rest of the body if the cell killed itself by apoptosis quickly before necrosis takes hold and explodes the cell, causing damage to neighbouring cells. However, apoptosis is not always quick enough to pre-empt necrosis. And if the damage is severe enough there will not be enough energy to power apoptosis.

As we heard earlier, the hara-kiri of apoptosis is executed by caspases. These are tiny proteins within the cell that once activated act like minuscule scissors, and go around the cell cutting up other protein molecules. The caspases are activated by being cut by other caspases, so once one caspase is activated the caspase cascade snowballs into an avalanche of molecular snipping, reminiscent of the finale of a slasher movie.

Damaged mitochondria can set this whole cascade going by releasing factors, normally kept safely within the mitochondria, which once free activate the caspases, thus setting the cell on course for death. In 1996 the new but rapidly expanding field of apoptosis research was agog when Xiaodong Wang from the University of Texas reported that the fatal avalanche was initiated by the release of cytochrome c from mitochondria. Cytochrome c is a small and venerable protein, known for over 60 years to be central to energy production in virtually all forms of life. What was it doing initiating cellular suicide? Nobody I talked to at the time believed it could be true. However, in this case, the unbelievable turned out to true. Cytochrome c is normally safely tucked away in the mitochondria, doing its vital job of generating electricity. But if the mitochondria are punctured, then cytochrome c is released from the mitochondria into the rest of the cell, where it activates the deadly caspases.

The caspases are the final executioners of apoptotic cell death, but the triggers for cell death are the holes or pores that puncture the mitochondria, and thus allow cytochrome c and other nasties out. One of these holes is known catchily as the 'mitochondrial permeability transition pore'. This pore is normally kept closed, otherwise the mitochondria just could not function to provide energy to the rest of the cell. But the pore is sensitive to lethal stimuli, such as free radicals or too much calcium, and in these conditions it opens wide. This has dire consequences for the cell, as firstly the mitochondria are incapable of producing energy in this condition, thus seemingly condemning the cell to necrotic death. And secondly the opening of the pore causes the mitochondria to release cytochrome c, which then activates the caspases and thus triggers apoptotic death. Thus pore opening is the trigger for a double dose of death.

This seems to be what happens to our cells during a heart attack or stroke. As we have heard before, a heart attack or stroke is caused by a blockage of the blood vessels supplying the heart or brain, so that the cells of these organs can't obtain sufficient oxygen to supply their mitochondria. Mitochondria depend on a continuous oxygen supply to burn the food we eat, and thus generate the electricity that powers our cells. They consume virtually all the oxygen we breath in and virtually all the food we digest to produce our energy, but in the absence of oxygen they can produce nothing. In these unhappy circumstances the mitochondria of heart and brain commit suicide by opening the pore, resulting in either necrotic or apoptotic cell death. Some cells die by necrosis (probably those cells that were most damaged by the blockage), while others die by apoptosis (and are rapidly cleared away). In the brain neurons are particularly dependent on mitochondria for the generation of their energy, so when the oxygen runs out, the neurons are rapidly de-energized. This results in the loss of the neuron's electrical field and the consequent release of glutamate, which then activates NMDA and other receptors on the neuron's surface. This now allows a wave of calcium to enter the neuron, which then hits the neuron's mitochondria, opening the permeability transition pore and triggering cell death.

Drug companies are now racing to develop drugs that block apoptosis or pore opening, which may be of benefit in a whole range of diseases. However, it is unclear whether blocking apoptosis would be beneficial, as a damaged cell may then be diverted to necrosis, causing further damage. Or even if necrosis is blocked, a damaged cell may function abnormally, causing delayed damage and dysfunction.

Perhaps our cells entered into a Faustian bargain when they first embraced the mitochondria two billion years ago. In return for the greatly increased capacity to produce energy, the cell also accepted a suspended death sentence, a kind of sword of Damocles hanging by a thread over the life of the cell. In fact, it has been suggested that the ancestors of our mitochondria that were embraced two billion years ago by larger cells already contained a toxin, which deterred the larger cell from destroying the proto-mitochondria living within it. Over time the larger cell, which became our cells, gained some control over the release of this toxin, and thus programmed cell death evolved. However, it is a dangerous game playing with death, and inevitably the game sometimes goes wrong, with fatal consequences.

We will encounter mitochondria again through this book as the sources of death and aging, yet paradoxically they are also the source of something essential to life: energy. In the *Star Wars* films of George Lucas, our mitochondria were myth-ologized as 'midichlorians', a symbiotic life form living within our cells, which generate the 'force'. But the 'force' has a dark side, just as our mitochondria do. If you watch the films, you will see just how close the midichlorians are to the real origin of our life force: the mitochondria.

Programmed death is common at many levels of biology. At the molecular level, most molecules in the cell are being continuously synthesized and degraded. For example, the proteins (which constitute the cell's machinery) are con-stantly being made within the cell by factories known as ribo-somes. And after a variable time (between minutes and days) the proteins are broken down by a cellular disassembly plant known as the proteosome. Consequently, no individual protein molecule within our cells lives more than a few days. And this is true for almost all other molecules in our cells. The exception is DNA, which although continuously damaged and repaired, does not normally undergo programmed destruction. Failure to degrade damaged molecules within our cells is thought to be a major cause of degenerative disease and aging. Failure to repair or destroy damaged DNA causes cancer. Failure to degrade fats damaged by free radicals in our blood vessels causes atherosclerosis. Failure to degrade aberrant proteins that stick together causes neurodegenerative disease.

At the sub-cellular level, most of the large structures of the cell, such as the mitochondria, are continuously being engulfed and degraded by a process known as autophagy (self-eating). Again, insufficient autophagy may allow damaged structures and molecules to build up within the cell, and contribute to neurodegeneration and aging. Autophagy is also used by the cell to adapt to new circumstances: if we fast or starve, our liver cells chew up much of their contents by autophagy in order to mobilize resources for the rest of the body.

At the cellular level, apoptosis and other forms of cell death contribute to the rapid turnover of our cells. The cellular lining of the gut is being continuously shed, digested and regrown. At a slower rate, the skin and most other linings within and outside the body are being continuously lost and replaced. Phagocytic white blood cells continually patrol the body, immediately eating any dead, damaged, aberrant or infected cells. We are in fact being incessantly eaten from the inside. Even the white and red blood cells are being continually replaced: an adult replaces 200,000,000,000 red cells per day, 10,000,000,000 white cells per day, and 400,000,000,000 platelets per day.

At the organ level, the linings of the womb, gut, lungs and skin undergo programmed destruction, which at the cellular level is executed by autophagy, apoptosis or simple shedding.

At the organism level, 'phenoptosis' may cause the programmed destruction of some individuals. As we shall see later in the book, even whole species may have a programmed lifetime. So at every level, our body is being continuously destroyed and rebuilt. Controlled death is intrinsic to life, but it is normally only when it gets out of control that it contributes to the death of the individual.

Although programmed death of organisms may occur in some cases and in specific circumstances, the general scientific consensus is that it does not have an important role in aging or in most deaths. However, the discovery of programmed cell death in the 1990s has reawakened interest in the idea that death may at least in specific circumstances have an adaptive value. Vladimir Skulachev, a former rector of Moscow State University, has named such programmed death of an organism 'phenoptosis'[1]. Examples of phenoptosis in action include bamboo, which lives for 15–20 years using vegetative multiplication, then comes into bloom and dies when the seeds ripen. The males of some squid and salmon species die just after transferring their sperm to the female. The bacterium *Escherichia coli* commits suicide if infected by a particular virus, thus curtailing viral multiplication and protecting the nearby *E. coli* population from infection.

Skulachev has suggested that sepsis (also known as blood poisoning) may play an analogous role in humans. That is, sepsis has evolved to rapidly kill off individual people with high levels of infectious bacteria in the blood, and thus protect the surrounding population, which is likely to include genetically related individuals. Sepsis is a common cause of death from wounds, surgery or infection generally, particularly in those people with a compromised immune system, such as the aged. Sepsis is the body's reaction to the appearance of large numbers of bacteria in the blood (septicaemia). The body's reaction includes inflammation, low blood pressure and organ failure. The patient dies due to the reaction rather than the bacteria themselves, which appears paradoxical. However, Skulachev argues that the rapid death in these circumstances may have adaptive value in preventing spread of the infectious bacteria to near relatives. If the individual is going to die anyway due to an out-of-control infection, it may be better to die quickly, before the infection is transferred to the children or other potential bearers or helpers of the genes.

Sepsis can kill bacteria-infected individuals of any age. More controversially, Skulachev has suggested that stroke, heart attack and cancer (the main killers in the developed world) are mechanisms of age-dependent phenoptosis. They are a means of killing off aged individuals who are no longer making a net contribution to their community. Kim Lewis has also supported the idea of programmed, age-dependent death of human individuals[2]. He postulated that in old humans a special mechanism of programmed death was switched on in individuals that were useless to others. Lewis wrote:

In humans, older individuals in early societies who are no longer useful could increase reproductive success by activating this programmed aging mechanism, which would result in channelling resources to progeny. Decreased emotional support and mastery are mortality risk factors in the elderly, supporting this hypothesis of programmed death in humans.

Indeed, there is known to be a correlation between mortality and psychological factors such as stress, lack of emotional support or lack of control over one's life[3]. Emotional stress can promote vascular disease, it can induce heart attacks, it can suppress the immune systems, and it may favour some forms of cancer. Lewis and Skulachev argue that aged humans who no longer contribute to the community become stressed in physical or emotional ways that promote death by stroke, heart attack or cancer. And at some stage in human evolution this benefited the community that was supporting their genes.

Skulachev proposed that there may be a link between phenoptosis (programmed death of the organism) and apoptosis (programmed death of the cell), in that too much or too little apoptosis can cause death of the individual organism. For example, during stroke or heart attack, apoptosis is triggered in brain and heart cells, and this cellular suicide may contribute to the pathology and subsequent death of the individual. In contrast, apoptosis may help to prevent cancer cells surviving, but a failure of apoptosis in the aged may promote cancer. So in theory, if apoptosis were turned down in the aged it would promote cancer, while if it were turned up in the brain and heart it might promote stroke- and heart attack-induced deaths. Thus in principle, evolution could use apoptosis to promote age-dependent phenoptosis. In practice, there is little evidence that apoptosis is deliberately activated (or inactivated) to kill organisms.

One consequence of programmed destruction at different levels of the body is that although we may appear stable, in fact we are continuously changing – most molecules in our bodies are replaced every few days. The form remains roughly the same, but the substance is changing. And even the form gradually shifts over time. The apparent stability of body and mind is an illusion.

The continuous destruction and resynthesis of our bodies consumes vast amounts of energy – that is the main reason that our adult bodies continue to consume energy even when we are at rest. Why resynthesize much of the body every few days? Partly it is done in order to change: to enable the cells and body to adapt according to circumstances and development, for example during infection or menstruation. Partly it is done in order to stay the same: to replace damaged molecules and cells, which would otherwise accumulate to cause dysfunction, aging and disease. Aging is at least in part caused by the accumulation of damaged molecules that cannot be replaced: mutated DNA, indigestible proteins and collagen fibres. So resynthesis and replacement are crucial to staying the same. But the frantic activity is also a source of damage and aging: the associated energy generation produces toxic free radicals, rapid protein synthesis results in misshaped proteins, and DNA replication generates mutations. These are the root causes of aging and degenerative diseases. The body can respond by replacing these molecules or cells faster, but this results in the whole futile cycle turning faster – running faster and faster in order to stay in the same place.

In this chapter we have seen that death is an integral, indeed essential, part of life. Our bodies and brains are changing continually, throughout life, from the moment of conception until, and beyond, the last breath. This suggests that a digital theory of life and an atomic theory of self are untenable. An ever greater challenge to these twin theories is provided by aging, the subject of the next chapter.

interlude 7
when do we become old?

All the world's a stage,
And all the men and women merely players:
They have their exits and their entrances;
And one man in his time plays many parts,
His acts being seven ages. At first the infant,
Mewling and puking in the nurse's arms.
Then the whining school-boy, with his satchel
And shining morning face, creeping like snail
Unwillingly to school. And then the lover,
Sighing like furnace, with a woeful ballad
Made to his mistress' eyebrow. Then a soldier,
Full of strange oaths, and bearded like the pard,
Jealous in honour, sudden and quick in quarrel,
Seeking the bubble reputation
Even in the cannon's mouth. And then the justice,
In fair round belly with good capon lined,
With eyes severe and beard of formal cut,
Full of wise saws and modern instances;
And so he plays his part. The sixth age shifts
Into the lean and slipper'd pantaloon,
With spectacles on nose and pouch on side,
His youthful hose, well saved, a world too wide
For his shrunk shank; and his big manly voice,
Turning again toward childish treble, pipes
And whistles in his sound. Last scene of all,
That ends this strange eventful history,
Is second childishness and mere oblivion,
Sans teeth, sans eyes, sans taste, sans every thing.

William Shakespeare, *As You Like It*

When do we become old? Above Shakespeare reflects the contemporary idea of the seven ages of man. This uses the Biblical idea that the maximum lifespan was 70 ('The days of our years are threescore years and ten', *Psalms* 90). And those 70 years were divided into seven ages that roughly correspond to decades. A man, if he survives, may play these seven parts successively, adopting different abilities, functions and characters according to the age. But he is fated to decline over the last decade or two of his life (50–70 years) into a second childhood. Thus aging, old age and death were somehow the mirror images of development, childhood and birth.

Fifty years after Shakespeare, John Milton wrote:

Death thou hast seen
In his first shape on man; but many shapes
Of death, and many are the ways that lead
To his grim cave, all dismal; yet to sense
More terrible at th' entrance than within.
Some, as thou saw'st, by violent stroke shall die,
By fire, flood and famine; by intemperance more ...
So may'st though live, till like ripe fruit thy drop
Into thy mother's lap, or be with ease
Gathered, not harshly plucked, for death mature:
This is old age; but then thou must outlive
Thy youth, thy strength, thy beuty, which will change
To withered weak and grey; thy senses then
Obtuse, all taste of pleasure must forgo,
To what thou hast, and for the air of youth
Hopeful and cheerful, in thy blood will reign
A melancholy damp of cold and dry
To weigh thy spirits down, and last consume
The balm of life.[1]

Milton here reflects the historical idea that aging is caused by the gradual consumption during life of some quality (innate moisture and heat), so that when this runs out, we are left cold and dry. This is connected to the more recent rate-of-living theory, which suggests that if 'you live fast, you die young', as we shall see in the next chapter.

Aging is not a modern invention: it's just a lot more common than it used to be. In the oldest extant medical book – the *Huang Ti Nei Ching Su Wen* (*The Yellow Emperor's Classic of Internal Medicine*), written in China some 3,500 years ago – the mythical emperor is being instructed about old age by the learned physician Chi Po, who tells him:

> When a man grows old his bones become dry and brittle like straw [*osteoporosis*], his flesh sags and there is much air within his thorax [*emphysema*], and pains within his stomach [*chronic indigestion*]; there is an uncomfortable feeling within his heart [*angina or the fluttering of a chronic arrhythmia*], the nape of his neck and the top of his shoulders are contracted, his body burns with fever [*frequent urinary tract infections*], his bones are stripped and laid bare of flesh [*loss of lean muscle mass*], and his eyes bulge and sag. When then the pulse of the liver [*right heart failure*] can be seen but the eye can no longer recognize a seam [*cataracts*], death will strike. The limit of a man's life can be perceived when a man can no longer overcome his diseases; then his time of death has arrived.

Nowadays we tend to regard the age of *retirement* as the watershed between adulthood and old age. But the state age of retirement is a relatively recent invention located at an arbitrary age[2]. Up until the Industrial Revolution, retirement at a particular age was an option open only to a relatively few rich Europeans. Everyone else worked until they dropped: either because they died or because they were no longer capable of working. The magic retirement age of 65 has been attributed to Germany's first Chancellor Bismarck, who in the 1880s, when pressed to pay pensions to war veterans, reluctantly agreed. When further pressed to set an age at which these veterans might receive their pensions, Bismarck is reputed to have demanded: 'How old are they when they die?'. His officials replied: 'Around 65'. And Bismarck retorted 'Then they get their pensions at 65!'. Bismarck himself did not retire from the Chancellorship until he was 75, and a very astute Chancellor he turned out to be. Germany was thus the first country to introduce a state pension in the 1880s, a model that was followed by most Western countries, although it was not until 1935 that the United States introduced old-age pensions to relieve hardship created by the Great Depression.

Nothing magical occurs at the age of 65 to start old age. Aging starts long before then, and it continues long afterwards. One definition of aging is the propensity to die by internal causes. Figure 2.2 illustrates your relative chance of dying at various ages: it is high at birth, and declines to a minimum at about age 10, but increases more or less exponentially thereafter. There is no absolute way to distinguish aging from development, or to separate growing up from growing old. We start aging the day we are born – or even before. Our cells start accumulating mutations and damage as soon as they exist: the aging process would appear to be an inevitable consequence of life. But why and how do we age?

8
aging

The evolution of death and aging

One of the most ancient and famous of riddles is that of the Sphinx: What goes on four legs in the morning, on two legs at noon, and on three legs in the evening? According to Greek mythology, Oedipus provided the correct answer: A man, who crawls on all fours as a baby, walks on two legs as an adult, and walks with a cane in old age. This answers the riddle of the sphinx, but it does not answer a second riddle that lurks within the first: Why is it that humans, having spent so much time developing and so much effort learning to become adults, then decay and revert into a kind of second childhood?

Whether death and aging have a meaning or purpose, whether there is rhyme or reason to these twin insults to human dignity, has much exercised our explanatory powers for at least the last few millennia. In the present age responsibility for spinning such explanations, at least in the biological realm, has been handed over to evolutionary biologists[1]. Evolution and death have an intimate relation, as evolution by natural selection would be difficult without death – not impossible, but certainly difficult. But aging and death present an interesting paradox for evolutionists, as they have no obvious evolutionary advantage, yet their universal existence cannot be denied. There have been four main evolutionary theories of aging proposed to solve the paradox: (1) the theory of programmed death, (2) the mutation accumulation theory, (3) the antagonistic pleiotropy theory of aging, and (4) the disposable soma theory. The last three theories are not mutually exclusive, and are often combined. Note that these are *not* theories of *how* aging occurs (which we shall encounter in the next section), but rather evolutionary explanations of *why* aging occurs.

The great German biologist August Weismann was one of the first and most prolific evolutionary theorists of death. In

1899 Weismann suggested that aging and death are an adaptive process: aging has been positively selected with the aim of limiting the overpopulation of the species, as well as to ensure genetic diversity. He wrote[2]:

> I consider that death is not a primary necessity, but that it has been secondarily acquired as an adaptation. I believe that life has been endowed with a fixed duration, not because it is contrary to its nature to be unlimited, but because the unlimited existence of individuals would be a luxury without any corresponding advantage.

The argument was basically that if the environment can only sustain a limited population, it is better that this population has a rapid turnover, so that it is genetically diverse and can rapidly adapt to new conditions. A species with a short lifespan is continuously changing its genetic makeup, and thus if the environment or competition changes, the species' genotype can be selected to match the new situation. Older individuals would age and die in order to make room for genetically superior progeny. In contrast, if a species is immortal, and resources are insufficient to sustain new individuals, then the species genotype is fixed and can not be selected to adapt to new conditions. In these conditions the short-lifespan species will out-compete the immortal species, leading to curtailment of 'immortality'.

In contrast to Weismann, twentieth century biologists came to the revolutionary conclusion that aging is not part of the design or genetic programme of our bodies. This is because aging is not an adaptive process – it does not benefit our genes. In 1952[3], Peter Medawar (who won the Nobel Prize in 1960 for research on immune rejection), brought together the ideas of previous authors such as J. B. S. Haldane, and postulated that aging was outside the sphere of action of natural selection – there was no selection for or against aging. This is partly due to the fact that the force of selection lessens as individuals advance in reproductive age. If an inherited mutation, such as that causing Huntington's disease, does not impair the individual until after several rounds of reproduction, then the mutation can spread in the population even though it is eventually deleterious to the individual concerned. Thus genetic changes that promote a disease or detrimental process that does not strike until after we have reproduced can spread simply because there is nothing to stop the genes being reproduced. So evolution by natural selection will not eliminate aging if aging occurs after reproduction.

However, there is another important reason for a lack of selection pressure on aging. In the wild, animals die by disease, starvation or predation long before they age. So there is no selection pressure to prevent aging or aging-related deaths. Aging is not the result of design, but rather of genetic drift that is not selected against. This allows DNA mutations to accumulate in the genome that have damaging effects only in old age. In the wild, animals would die due to external causes long before these mutations could cause trouble. So natural selection could not act to prevent or repair such mutations. Very few animals survive to old age in the wild, because they are struck down by disease, starvation or predation long before they have any chance of suffering from old age. Whereas in artificial, protected, zoo or laboratory conditions, where external causes of death are minimized, animals reach old age, at which point internal causes of death dominate.

In Medawar's words, aging and degenerative diseases have been revealed 'only by the most unnatural experiment of prolonging an animal's life by sheltering it from the hazards of its ordinary existence'. Because selection of animal and our own genomes occurred (up until a few thousand years ago) in wild rather than protected conditions, no selection pressure has been exerted on aging, as in wild conditions no one survived long enough to age. The vast majority of random mutations accumulating in genomes are neutral or detrimental, rather than promoting survival, so that if no selection pressure is exerted on old age, then detrimental mutations affecting function and survival in old age will inevitably accumulate.

In 1957[4], George Williams argued that even though animals in the wild may not die of old age and degenerative diseases, still the progressive decrease in motor, sensory, reproductive and immune functions with age would have a substantial impact on mortality relative to competing younger animals. For example, an older animal that was not as fast as its younger kin would be more likely to die of starvation or predation. Thus we still needed some explanation of why aging was not eliminated by evolution.

Williams proposed that genes favouring aging could be maintained during evolution due to their 'antagonistic pleiotropy'. By this he meant that genes which had a beneficial effect when the individual was young might have an unintended detrimental effect when the individual was old. Their beneficial effects on the young would cause them to be positively selected for, while their detrimental effects in the old would cause them to be selected against, but the former selection would be more powerful. As a result, senescence would be no more than a side effect of natural selection. The detrimental action of the gene might be an accidental side-effect of the beneficial action. For example, a gene that promoted rapid cell growth and division would be beneficial in the young, but the same gene might promote cancer in the old. Or a gene may mediate an important process (such as energy production), but with a toxic side reaction (such as free radical production), which caused accumulating damage, insignificant in the young, but significant in the old.

The last of our evolutionary theories of aging is the disposable soma theory, proposed in the 1970s by Tom Kirkwood[5] (Professor of Medicine at the University of Newcastle's Institute for Aging and Health), based in part on some late ideas of Weismann. Kirkwood distinguished between our reproductive cells (the germ cells: the sperm and eggs) and the non-reproductive, somatic cells of the rest of the body, or soma. He highlighted three fundamental points. First, the only purpose of somatic cells is to transmit genes to the following generation, which is why an organism's soma can be considered 'disposable'. Second, reproduction has a high energy cost, as do the repair and maintenance mechanisms of cellular functions. These mechanisms are absolutely necessary, since a certain amount of cellular deterioration is inevitable. Third, energy resources are scarce and difficult to obtain, so reproduction and cell maintenance therefore compete for energy use.

Since, even with efficient maintenance systems, organisms die from external causes (hunger, predators, disease and accidents), it is more profitable to prioritize energy investment in reproduction, albeit at the expense of less efficient maintenance systems that lead the individual to aging and death. Evolution is in the unhappy position of having a limited energy budget, and it must choose between investing in the soma or the germ line. But its shareholders are only interested in the bottom line: the total number of surviving genes. So evolution only invests in preventing soma aging and death to the extent that this promotes survival of the genes. The theory does not explain why we age, but rather why evolution does not eliminate aging. That is because it is better to use the limited resources available on reproduction rather than on preventing aging, because the individual is going to die anyway before it ages. Aging and death are the price we pay for sex and the immortality of our genes.

This depressing theory echoes those of Aristotle, 23 centuries earlier: sex is responsible for mortality and that each sex act has a direct life-shortening effect. This troubling concept enjoyed remarkable longevity, and was clearly expressed in the seventeenth century writings of Metaphysical poets such as John Donne: '...since each such act, they say, Diminisheth the length of life a day, ... I'll no more dote and run, To pursue things which had endamaged me'. There is some experimental evidence for this idea in fruit flies, in that unmated flies live longer, and a component of the seminal fluid shortens the life of mated females. However, reassuringly, there is no supporting evidence in humans.

The hypothetical trade-off between investing in reproduction or investing in aging prevention has been used to explain the different lifespans of different species. Small animals like mice die young because they have many predators, and thus it is pointless to invest in preventing aging. During their evolution, large animals like elephants and humans have had relatively few serious predators, and thus natural selection has indeed invested in extending lifespan. Humans have remarkably long spans relative to other animals. This may be good news for us now, but it may make it relatively difficult in the future to further extend our years. Evolution has already spent millions of years fiddling with our genes to extend human lifespan. So it is being very optimistic to believe we can substantially further extend human lifespan beyond that of any other mammal.

Is there a trade-off between fertility and aging in humans? A thousand years of records of the British aristocracy indicate that those reaching a ripe old age (over 80) had reduced fertility (fewer children, produced later in life). One possible mechanism of this trade-off is suggested by the finding in humans that reduced fertility is associated with a highly active immune system. The immune system needs to be turned down during pregnancy in order to prevent rejection of the foetus, but a docile immune system leaves the mother susceptible to infections[6].

It is important then to distinguish between the evolution of aging and the evolution of death, because, during most of animal and human evolution, aging did not significantly contribute to death. So while natural selection could not act directly on aging, it certainly did act on our susceptibility to the main agents of death: infection, predation and starvation. And huge amounts of energy, time and other resources are invested in preventing these, for example the effort expended on the immune, nervous and muscular systems of the body. Wild animals spend large amounts of time and energy acquiring food, avoiding becoming food, and fighting infection. But this time and energy, invested in preventing death, takes away from the time and energy that could be invested in reproduction. Evolution must seek the optimum balance between sex and death. And that balance will determine the average lifespan for the species in the wild. In this sense, then, evolution determines death, but only because there are limited resources to prevent death.

While Medawar, Williams and Kirkwood assert that aging was not selected for by evolution, Weismann claimed that aging is a strategy that has been designed by nature. Weismann's theory was subsequently rejected by evolutionary biologists for two principal reasons. Firstly, as we have already heard, most animals in the wild die due to external causes before aging begins. An aging strategy is therefore not necessary to control overpopulation. Moreover, it is difficult to accept that a strategy that is so infrequently put into practise in nature could be actively selected. Secondly, it is now accepted that natural selection acts at the level of individuals and their genes, not at the level of the species. If aging were beneficial, for example by promoting genetic diversity, the supposed benefit would be at the level of species, not the individual. Aging does not offer any advantage to an individual organism, since it increases its chances of dying and reduces its reproductive capacity. So genes that promote death and aging have no chance of spreading through the population. Thus for more than half a century the dominant opinion has been that aging is a non-adaptive phenomenon. This leads to a stark picture of the biological function of aging: there is none[7].

Tom Kirkwood has recently[8] summarized the current thinking on aging:

> One of the strangest things about the aging process is that, despite its near universality among higher organisms, it is something of an artefact. In the wild, aged organisms are extremely rare because most animals die young. Because old age in nature is a rarity, any idea that the aging process has been actively favoured through natural selection – such as by evolving 'death genes' – to keep population size under check, is almost certainly false. Put simply, we are not programmed to die.

What causes aging?

Having considered *why* we age, we now need to know *how* we age. What is the mechanism by which aging is caused[1,9,10]? Unfortunately, aging is a frontier area of research, a kind of Wild West of science, because there is no mature theory of what causes aging. There are lots of ideas (at least 300 theories of aging have been proposed), but no one really knows. We are not even sure what aging *is*. There is no universally accepted definition of aging. It may be several different processes, and it may have many different causes. There may never be one unified theory of aging.

One common definition of aging is a progressive decline in body functions. In humans, aging is accompanied by decreased motility, increased response times, lower sensory acuity, and poorer memory. We also get an increased incidence of prostate or breast cancer, high blood pressure, skin wrinkling, osteopoposis, cataracts etc. Another common definition of aging is an increased propensity to die. In the early nineteenth century, an English actuary, Benjamin Gompertz, discovered that after the age of 30, your chances of dying double every eight years (Figure 2.2). This law-like relation was the founding stone of the life insurance business. And it suggests that you have no chance of cheating death (and the life insurance business), because the longer you live the more likely you are to die.

The doubling of death rates every eight years seems to be relatively independent of conditions affecting the actual death rate. So although death rates at any age are lower now that in the early nineteenth century, the rates still double every eight years. Indeed, the same doubling time appears to have held in Japanese-run prisoner-of-war camps, although the death rates at any age were 10 times higher. The same doubling time may even have held in the Stone Age when people died at roughly 150 times the present rate[10]. In the present day, men have higher death rates at all ages than women, but the mortality doubling time is the same. However, different species have different doubling times, indicating different rates of aging. For example the mortality doubling time of mice is three months, and that of fruit flies about 10 days. Thus some scientists prefer to measure or define the rate of aging as the mortality doubling time.

The law-like nature of the Gompertz relation also suggests that the aging body is following some intrinsic, fixed programme, just as the body follows a fixed programme to develop from egg to adult. But in contradiction to this, the details of aging are very variable for different people. We all age in different ways and at different rates, suffering from different diseases and dying from different causes. This suggests (in contrast to development that follows a fixed path) that aging does not follow a fixed programme, but rather consists of a random accumulation of faults.

Why does aging occur? The general idea is that as we age our cells gradually accumulate a host of minor faults and damage that cause the age-related decline in function of cells, organs and body systems. The molecular accidents that cause these faults are not designed by the biology; rather, they just occur by accident in a random and unpredictable way. And as these faults accumulate with age, eventually they precipitate organ dysfunction, disease or death. The same thing happens with a car or computer – it is pristine when new, but it gradually loses its gloss and accumulates faults and corrosion, which eventually precipitate automotive or silicon death.

Most forms of damage in the cell are potentially reversible, but some forms of damage to the DNA (mutations) are not easily reversible, and so accumulate with age. Support for a DNA damage theory of aging comes from the finding that genetic diseases associated with accelerated aging, such as Werner syndrome, are associated with high rates of DNA mutation. This suggests that DNA damage and mutations can cause aging. However, although DNA damage does accumulate with age in humans, the amount of damage is generally very low, so that it is still unclear whether this low level of damage could be the cause of significant dysfunction.

Although mutations are relatively rare, aneuploidy, the occurrence of cells with greater or less than the correct number of chromosomes, is relatively common during aging. All of our DNA is packed into 23 pairs of chromosomes, and each of these needs to be replicated before each cell division. But if things are too lax or frantic, one or more chromosomes may end up in the wrong daughter cell during cell division, so that instead of having a pair of a particular chromosome, a cell may have 1, 3 or 4 versions of the same chromosome. The frequency of aneuploid cells in human females increases from 3% at age 10 to 13% at age 70. This may cause problems, at least with fertility: 1% of embryos from women in their 20s and early 30s are aneuploid, but in women older than 40 years the figure is over 50%. Most aneuploid embryos die early in pregnancy, so this contributes to the loss of women's fertility with age. The other major factor is that women are born with a fixed number of eggs, and the supply eventually runs out.

However, if aneuploid embryos are not eliminated during development in the womb, this can cause trouble too. Children born with one extra copy of chromosome 21 in each of their cells have Down's syndrome. This is the one and only cause of Down's syndrome, and the risk of producing a Down's syndrome baby increases dramatically with the age of the mother, because of the increased frequency of aneuploid embryos. You might think that an extra copy of a chromosome is a useful thing to have on a rainy day. But it is almost always detrimental. In the case of Down's syndrome, the extra chromosome causes developmental problems, leading to malformed organs and mental retardation. In addition, it causes accelerated aging, with premature onset of degenerative diseases, resulting in a shorter life expectancy (55 years).

A person with Down's syndrome has aneuploidy in 100% of their cells. Although the rest of us have a much lower level of aneuploidy at birth, this percentage of aneuploid cells increases with age. And this, together with other accumulating damage to our genomes, may contribute to aging.

The theory that aging is due to accumulating damage to nuclear DNA was dealt a potentially fatal blow by the creation of Dolly, the first cloned sheep. Dolly was created by taking a cell from an adult 'donor' sheep, removing the nucleus (which contains the DNA), and injecting this nucleus into a newly fertilized egg from which the normal nucleus had been removed. If the accumulating DNA damage theory was correct, then the donor nucleus would be aged, so that Dolly would already have had an adult level of aging before she was born. As this was not immediately apparent in Dolly, it suggests that there is something wrong with the theory that aging is written in the nuclear DNA. However, it is not entirely conclusive, because Dolly eventually did appear to suffer from accelerated aging, and had to be killed because she had serious arthritis at what was a premature age for sheep. Furthermore, cloning is a notoriously inefficient process, requiring hundreds of eggs to yield one live birth. Because aging is a statistical process, it is possible that some donor nuclei are relatively free from aging damage, and it is these that produce viable, relatively unaged embryos and clones.

If aging is caused by an accumulation of damage (to the DNA or elsewhere in the cell), what is causing the damage? One of the most popular theories is the 'free radical theory of aging', devised by Denham Harman[11] in 1956. According to this theory the body is continuously producing free radicals as a by-product of its normal function. These free radicals directly attack DNA, proteins and fat to cause the damage that accumulates with age, and ultimately causes aging. The name 'free radicals' may suggest a benign group of liberal revolutionaries, but in fact they are a vicious bunch of toxic chemicals. They result from the leakage of electrons out of the cell's machinery: one electron inadvertently added to oxygen produces superoxide, two electrons leaked to oxygen produces hydrogen peroxide, and three electrons produces the most toxic of the bunch: the hydroxyl radical. Superoxide, hydrogen peroxide and the hydroxyl radical are the most famous free radicals, although hydrogen peroxide is only an honorary member of the club as it lacks a radical (a free unpaired electron), and there are a number of less famous members.

Evidence that free radicals were indeed bad and the body wanted to get rid of them came with the discovery of a protein found in all our cells whose sole function was to dispose of superoxide – it was named 'superoxide dismutase'. Further support for the free radical theory came from experiments on mice, where the gene encoding superoxide dismutase was disrupted or 'knocked out'. These mice lacked part of the cellular machinery that inactivates free radicals, and therefore had higher levels of free radicals and were found to die young, apparently due to more rapid aging. This suggests that increasing the amount of free radicals might increase the rate of aging. But does decreasing the amount of free radicals decrease the rate of aging?

This was tested by feeding antioxidants to the nematode worm *C. elegans* and the fruit fly *Drosophila*. Antioxidants are a wide range of chemicals that mop up damaging oxidants such as free radicals, and thus should decrease the level of damaging free radicals in the body. Feeding antioxidants to nematode worms and fruit flies did increase their lifespans, suggesting that free radicals can indeed shorten life. However, these two species have very short lifetimes, counted in days or weeks, and recent experiments suggest that even in these species antioxidants can only extend lifespan in specific 'stressful' conditions. Similar experiments with feeding antioxidants to mice have failed to extend their lifespan. This suggests either that free radicals do not contribute to aging in mice or that the radicals are not efficiently removed by the particular antioxidants used.

A decade after his original hypothesis Harman went on to refine his theory and suggest that the main source and target of free radicals during aging was the mitochondria: hence the 'mitochondrial free radical theory of aging'. Mitochondria are the tiny, wiggling worms within our cells that generate our energy and control our death. Mitochondria produce all the energy and heat we need by burning the food we eat using the oxygen we breathe in. In fact, they consume almost all the food we eat and the oxygen we breathe in, and they produce almost all the energy we use and heat we need. To do this they rip electrons off the food molecules and feed them down an electron transport chain, basically a wire made of iron and copper atoms. And at the end of the chain the electrons are given to oxygen, reducing it to water. This generates an electric circuit that is used to create further energy for the cell. Thus, surprisingly, we are actually powered by electricity. We don't notice this because the electrical circuits are so incredibly small, wrapped as they are in the membranes of the mitochondria.

Unfortunately not everything works perfectly in the mitochondria. About 1% of the electrons leak out of the electron transport chain to react directly with oxygen to produce superoxide and other free radicals. This is one of the main sources of free radicals in the cell. That's why Harman suggested that mitochondria are one of the major sources of aging. Furthermore, mitochondria are particularly susceptible to free radical damage. Mitochondrial membranes, proteins and DNA are all relatively easily damaged by free radicals, and all of these potential targets do accumulate damage with aging. Mitochondrial function generally declines with age, resulting in the production of less energy, which might in turn contribute to the decline in tissue function with age. However, no one is really sure whether the decline in mitochondrial function is a consequence or a cause of aging.

Circumstantial evidence supporting a mitochondrial free radical theory of aging comes from genetic experiments on mice. Geneticists knocked out a key gene coding for a protein protecting our cells from free radicals: superoxide dismutase

(SOD), which destroys the free radical superoxide. But SOD comes in two versions: one located in the mitochondria and thus protecting the mitochondria from mitochondrial free radicals, and the other located in the rest of the cell. When the mitochondrial version was knocked out the mice died much earlier than when the non-mitochondrial version was knocked out. This suggests that mitochondrial free radical production and damage is more important for aging than that in the rest of the cell. But this work does not identify what target within the mitochondria might be important for aging.

One potentially important target is mitochondrial DNA, which can be mutated by free radicals. Recently, Nils-Goran Larsson of the Karolinska Institute in Stockholm, Sweden, set out to test whether mitochondrial DNA mutations could cause aging. His team created mice with a defective version of the mitochondrial DNA polymerase. This enzyme normally makes new copies of the mitochondrial DNA, but also repairs mismatched copies, i.e. mutations. The defective version of the enzyme was less capable of repairing mutations, resulting in mice that acquired mitochondrial DNA mutations two or three times as fast as normal. The question was, do these mice care? Early on in their lives, this change had little effect: the mice grew normally and seemed healthy. But later on things started to go wrong. At 25 weeks of age, they began to show signs of aging: baldness, spine curvature, osteoporosis, fertility loss and heart muscle disease, signs that do not normally appear in mice until they are least one year old. Lifespan was also dramatically reduced: normal mice live two to three years, but half the mutant mice had died by 48 weeks, and all of them were dead by 61 weeks[12]. So speeding up mitochondrial mutations can increase the rate of aging, at least in mice.

Perhaps a more important question for us lovers of life is whether slowing down the rate of mitochondrial mutations could slow down the rate of aging, but this is a more difficult thing to do. At the moment the only feasible way to attempt this is to try to reduce free radical levels in the mitochondria. Michael Murphy in Cambridge, UK, is trying to do this with mitochondria-targeted antioxidants, but we don't know yet whether this will have an impact on aging.

If less SOD can speed up aging you might expect that more SOD could slow it down. However, aging ain't that simple. Indeed, if it were we probably would have more SOD, because in evolutionary terms it is very easy to increase the amount of a protein. People with Down's syndrome have an extra copy of chromosome 21, which includes the gene for SOD, so their cells end up with more SOD than normal. However, instead of slowing aging, people with Down's syndrome age more rapidly. The reason might be that SOD, while getting rid of one free radical – superoxide – produces another: hydrogen peroxide. Furthermore, the hydrogen peroxide may react with SOD to produce even more damaging radicals, such as the nasty hydroxyl radical. So even though SOD is essential to get rid of one radical, in some conditions it can be damaging by producing other radicals. Some types of motor neuron disease are caused by mutations of the SOD gene. This results in increased production of these other radicals, which may progressively kill motor neurons. So, at least in the case of SOD, you really can have too much of a good thing.

It may seem paradoxical to consider aging as on the one hand a random, undesigned process, but on the other hand as a process determined by our genes. However, as explained in the last section, genes may produce detrimental aging late in life as a side-effect of beneficial processes promoted early in life. That genes influence aging seems undeniable. For example, the maximum lifespan of different organisms, with different genomes, covers an enormous range. The adult mayfly burns out its life in a day, while the gnarled bristlecone pine lives thousands of years. And animals, such as fruit flies, can be bred and selected to have longer lives, so that they inherit lon-gevity. But the inheritability of longevity is relatively small in humans. Only about 20–30% of the variability in longevity between people is genetic, the rest being due to environmen-tal factors (e.g. whether you smoke or not). That doesn't mean genes are not important, but rather that environment is more important.

Larger animals tend to live longer than smaller animals, so a mouse lives a maximum of three years, while an elephant lives for up to 60 years. Large mammals use energy more slowly (per gram) but live longer, so that the total amount of energy used in an average lifetime is roughly equal for all mammals (when expressed per gram). There are a number of exceptions to this general rule, including humans, who live up to four times longer than we should according to our body size. But overall, for all species of mammals, and many other animals, the relationship between maximum lifespan, body size and metabolic rate (rate of energy use) is impressive. Basically, ani-mals that live fast, die young. Any serious theory of aging needs to account for the fact that small animals with high met-abolic rates have much shorter maximum lifespans than large animals with low metabolic rates (per gram). The general theory that a fast pace of life (or energy use) somehow causes rapid aging is known as the 'rate-of-living' theory of aging.

The rate-of-living theory has a long history, and one of its most enthusiastic supporters and the man who gave it its catchy name was Raymond Pearl. Pearl (1879–1940) was a prolific scientist and popularizer of science at Johns Hopkins University, Baltimore, in the USA. Pearl, being unusually tall and intelligent, towered over his peers, both literally and figu-ratively. Pearl was one of the founders of biometry, the appli-cation of statistics to biology and medicine. In his lifetime, he published 17 books and more than 700 scientific papers; he also wrote for newspapers and literary journals, on a vast range of subjects from fruit flies to the link between smoking and cancer. With all this frenetic activity we might expect him to have died young, but in fact he reached a respectable age of 61, significantly older than the age, 50, at which he thought people became too foolish or senile to vote.

Pearl believed that aging was an inevitable side effect of a fast metabolic rate. The metabolic rate is the rate of energy production, measured by the heat produced or oxygen consumed by the body, and it determines the amount of food consumed and energy available. In fact he wrote an article for the Baltimore Sun in 1927 headlined 'Why Lazy People Live the Longest'. And he even attributed the greater longevity of women, relative to men, to the fact that they supposedly performed less physical labour!

Pearl collected data on the longevity of people in different occupations and professions with different amounts of physical labour and thus different metabolic rates. He found, perhaps unsurprisingly, that people working in occupations with high physical labour, such as miners, had on average shorter lives than those doing little physical work, for example academics such as Pearl. Of course, there may be many other explanations of this finding, including the effects of poverty, nutrition and health care, but Pearl thought it was strong evidence for the rate-of-living theory of aging. If this theory were true it would have the rather startling implication that exercise would shorten your life, while being a couch potato would prolong it. Unhappily (or happily for the puritans among us) we now know that this is not true. Professional athletes, who use considerably more energy on average, live just as long as the rest of us, but not longer.

Exercise may cause some free-radical damage in parts of the body, but this is offset by a healthier cardiovascular system. The overall effect of regular exercise on longevity and aging is small and unclear. Studies in laboratory rats (who may not be the best models of humans) have shown that exercise increases or decreases average longevity by a small amount, but has no significant effect on maximum age or aging. The largest human study, on the alumni of Harvard University, concluded that regular moderate-to-vigorous exercise (regular jogging for example) could increase longevity by one to two years, roughly the time taken to do the exercise over a lifetime. However, over-exercise or inappropriate exercise may shorten life slightly. Overall, exercise is thought to give some protection against diseases of middle age, but has no significant effect on aging or diseases of old age[13]. On the other hand, this information should not cause you to abandon the gym or hang up your running shoes. Although exercise does not lengthen life, it does make those years lived more healthy, and that is what really matters.

The rate-of-living theory of aging is dead, but we need some explanation of why the rate-of-living, measured by metabolic rate, is related to lifespan. Many aging researchers now believe that aging has something to do with accumulating damage due to free radicals, and that animals with a higher metabolic rate may produce more free radicals as a side product of their metabolism, and thus have a shorter lifespan. Basically, in order to produce more energy an animal needs more mitochondria. And more mitochondria result in the production of more free radicals, as up to 1% of the electrons used by mitochondria leak out to produce these toxic radicals. Thus the amount of free radicals produced by an animal may be roughly proportional to the amount of energy produced. So if an animal produces and uses a lot of energy it also produces a lot of free radicals, and this may shorten its life. However, we don't know that this is the correct explanation. It may well be that animals that live fast do everything fast, including all the processes that cause aging, so obviously they age fast.

There are other theories of aging that attribute a more active role to the body in causing aging. One of these is 'the telomere shortening theory of aging'. Telomeres are stretches of DNA at the ends of chromosomes that aid in the replication of the chromosome. Each time a cell divides, its chromosomes are replicated, and because of the way the DNA is duplicated the telomeres are shortened by a small amount. Eventually after many rounds of division the telomeres become so short that the chromosomes cannot be replicated. At that point the cell can no longer divide and is said to be senescent. This is known as the Hayflick limit of replicative senescence, after its discoverer. Leonard Hayflick discovered that cells grown outside the body, in a culture dish, replicated themselves quite happily up a certain number of cell divisions, and then stopped dividing. What is the purpose of this limit on cell division? It may have a role in preventing cancer, as cancer cells must divide very many times if they are to cause a nuisance. By setting a limit to the number of times a body cell can replicate, telomeres may limit the growth and therefore the impact of a cancer.

If the shortening of telomeres limits the total number of cell divisions in the body, could it contribute to aging of the body? Well, it takes roughly 40 cell divisions to produce the adult body from the fertilized egg, whereas the Hayflick limit for human cells is about 50 cell divisions, and most cells divide very little once we have become adults. For example, neurons, muscle and heart cells do not divide at all in the adult, and yet they age and contribute to the overall aging of humans. Thus, telomere shortening certainly can't contribute to the bulk of aging.

However, some cells in the body do continue to divide in the adult. For example, cells of the immune system need to divide to function, and telomere shortening in these cells might limit immunity in the aged. Also, endothelial cells that line the

insides of blood vessels divide rapidly when vessels are damaged or new ones are needed. Endothelial cells can be washed
off the sides of the vessel wall by the turbulent flow of blood,
particularly at vessel junctions. These cells must be replaced by
the division of the remaining endothelial cells, and it has been
found that the cells at these turbulent points have particularly
short telomeres, suggesting both that they have undergone
many divisions and that they may be close to senescence. It is
these very cells, the endothelial cells in turbulent vessels,
which are thought to be the origin of arteriosclerosis, which in
turn causes the age-related deaths of almost half the human
population.

So telomere shortening might contribute to some aspects of
aging, but this is far from proven. There is no direct evidence
that endothelial or immune cells do become senescent in our
bodies rather than a laboratory dish. There may in fact be a
link between the free radical theory and telomere theory of
aging, because free radicals can damage telomeres, resulting
in their shortening, and thus potentially speeding aging.

Telomere shortening is usually thought of as a means devised by evolution to prevent cancer, by setting a maximum
number of division for body cells. However, recently it has
been suggested that telomere shortening may contribute to
the generation of some cancers. As pre-cancerous cells divide
rapidly they whittle away their telomeres. These short telomeres cause chromosome instability: bits of chromosomes
stick onto or insert into other chromosomes. The result is
genetic havoc, which will often lead to cell death. But in a crucial small proportion of cases it may result in the activation or
inactivation of genes that promote cancer.

One of those genes is telomerase, which if activated can re-
lengthen the telomeres, thus restarting rapid cell division.
Telomerase is normally only active in germ cells such as sperm,
enabling them to avoid senescence and be potentially immor-

tal. Tumour cells can subvert the same trick in their own bid for immortality. So, according to this theory, evolution of a cancer within the body may go through a crisis when the telomeres shorten sufficiently. And this crisis will either stop the cancer developing if the crisis is not resolved, or the consequent genetic instability may greatly speed up cancer evolution. So the cancerous cells emerge from the crisis with a much more malignant and dangerous genome. Thus telomere shortening could, theoretically, contribute to both cancer and atherosclerosis, the main causes of age-related disease. If true, a dose of telomerase-activating drug or telomerase gene therapy could be just what the doctor ordered, or even what your junk email may be offering you next week. However, the scientific jury is still out on whether telomere shortening significantly contributes to aging and age-related diseases[13].

Recently the startling idea was suggested that telomere shortening may regulate the aging and 'lifetime' of entire species during evolution. Many species have come and gone during the history of the world. For example, the modern human species is supposed to have arisen about 100,000 years ago in Africa, while Neanderthals became extinct about 30,000 years ago in Europe. Reinhard Stindl in Vienna hypothesized that whole species may age due to the shortening of the telomeres of germ cells each time an animal replicates. A new species is supposed to start with long telomeres but after thousands or millions of generations they become so short that the individuals can no longer function or replicate properly. The species may then become extinct, or possibly pass through an evolutionary bottleneck where the telomeres are re-lengthened, and a new species born. However, while the idea of species aging is attractive, it currently has little scientific support or evolutionary rationale.

The only treatment currently known to substantially slow the rate of aging and increase maximum lifespan in mammals is calorie restriction. That means permanently reducing the food calories in the diet, usually by about one third of the normal intake. That's enough to make you permanently hungry, but it extends lifespan in mice and rats by up to 50%. This was first discovered by Clive McCay and his team at Cornell University, who fed rats from birth the normal level of nutrients, but reduced their calorie intake. The result was increased life expectancy of up to 100%. This experiment has been repeated in many different animals, including monkeys, with similar results. The calorie-restricted animals appear to be healthier too, as the rate of aging appears to have slowed down.

We don't know whether calorie restriction works in humans, but that has not stopped a lot of Americans trying. Roy Walford was one of the pioneers of human calorie restriction. He was Professor of Pathology at the University of California at Los Angeles, and spent two years as crew physician in Biosphere 2, the experimental, closed (bubble) ecosytem in Arizona. When Biosphere 2 became overrun with weeds, and the crew had to live on much reduced rations, Walford's knowledge of calorie-restricted diets came in very useful. He subsequently developed a range of calorie-restricted diets that he published in his book *Beyond the 120 Year Diet*. These became the inspiration for a Calorie Restriction Society, and a worldwide movement of frighteningly thin people. But Walford's death from motor neuron disease in 2004 at 79 years, despite religiously adhering to his own diet, was a great disappointment to the movement. However, the experimental work on animals strongly suggests that if a calorie-restricted diet is adopted from a young enough age, it will slow aging – at least in animals.

Why calorie restriction works is unclear. It may be because it reduces the amount of free radicals produced by the mitochondria. Fewer food calories means fewer electrons being fed into our mitochondria, and thus fewer electrons potentially leaking out to produce free radicals. Alternatively, it may be because fewer food calories results in lower blood levels of the hormones insulin and insulin-like growth factor 1. These hormones have short-term benefits, but if chronically elevated can cause long-term damage to the body. Mutations in genes related to these hormones have been found to extend lifespan in yeast, worms, flies and mice. These hormones can up-regulate growth in conditions of plenty, but they may also downregulate the body's defences against stress. For example, they may reduce levels of proteins such as SOD defending against oxidative stress. So starvation might extend life by changing hormone levels that regulate antioxidant and other stress defences. Or it may work by something different that we haven't even thought of yet. However it works, in the current cultural climate, permanent calorie restriction is unlikely to be a very practical or popular means of slowing aging. Indeed, the current obesity epidemic, which is affecting virtually the whole world, is thought to be one of the greatest threats to increasing lifespan in the twenty-first century[14].

At least 300 million people in the world are currently obese, and this is rising rapidly. In the UK 22% of adults are obese, and this level has tripled in the past 20 years. In the US, 30% of adults are obese. The weight of the average American increased by 4.5 kg (10 lb) during the 1990s. Obesity was responsible for $100 billion in medical costs and 300,000 deaths annually in the US. Obesity may, in contrast to calorie restriction, increase the rate of aging. It dramatically increases your chances of getting the age-related degenerative diseases arteriosclerosis, hypertension, heart disease, stroke and some forms of cancer.

To take just one example: obesity dramatically increases your chances of getting adult-onset diabetes. So the epidemic in obesity is causing an epidemic in diabetes. The estimated lifetime risk of developing diabetes for individuals born in

2000 in the US is about 35%. That is, one third of the US population may end up with diabetes – a condition which itself
can accelerate aging. Obesity, caused by calorie over-con-
sumption, may stimulate the rate at which our cells age. A
recent study by Tim Spector and colleagues[15] showed that
obesity speeded the rate at which our telomeres shortened, so
that obese people had shorter telomeres equivalent to an
extra nine years of aging. In part this may be due to increased
free radical production as well as the toxic effects of glucose
and fat themselves. Obesity can also cause fatigue, breathlessness and damage to bones. Obesity is likely to contribute to
the general increase in chronic, age-related disability and
degenerative disease in the future. But it's not all doom and
gloom: on the up-side, it could be fun getting obese!

A life in flames

There is a dawning realization among scientists that a single
process may underlie all age-related diseases, and that process
is not aging, but rather inflammation. This revelation may
have consequences not only for the treatment of age-related
disease, but also our concept of what and why these diseases
are. Inflammation is recognized by its four cardinal signs: calor
(heat), dolor (pain), rubor (red colour) and tumour (swelling),
defined by the Roman encyclopaedist Celsus in about 50 AD. It
occurs whenever you cut yourself or become infected. The
surrounding tissue becomes swollen, warm, red and painful.
And the damaged tissue is then invaded by various types of
white blood cell, which may appear as pus. The white blood
cells are there to clear up the mess and kill bacteria or other
nasty pathogens by releasing a cocktail of toxins, including
vast amounts of free radicals. However, if these cells are activated for too long (as in 'chronic' inflammation) these toxins

will also damage or kill body cells. This may be the origin of the diseases of aging, and perhaps even aging itself.

The body is armed with two types of immune system that defend against attack by pathogens. The innate immune system is the system that you and everyone else are born with. It detects pathogens using a small set of pattern recognition molecules that can bind to and recognize bits of the most common types of pathogen: bacteria, viruses, and fungi. It can also recognize damaged cells and tissues in the absence of pathogens. In contrast, the adaptive immune system acquires immunity against very specific pathogens and molecules during life. If we get infected by a new pathogen during life, this system generates a specific means of recognizing that pathogen (a specific antibody) that both helps attack that pathogen and enables the body to remember the pathogen.

The innate immune system is described as 'strong and stupid' while the adaptive immune system is 'lazy and smart'. The innate system is mobilized within a few minutes of detecting a problem, whereas the adaptive system takes one to two weeks to generate an effective response to a threat. That's why a cold or flu can drag on for so long. The four signs of inflammation are due to the innate immune system. The warmth and redness are due to blood being diverted to the damaged area. The swelling is due to the blood vessels becoming leaky so that the white blood cells can leave the blood and enter the tissue to attack the pathogens or damage. The pain is due to the release of chemicals from damaged tissue that stimulate pain receptors, alerting the brain to the problem.

Inflammation used to be thought rather boring. But within the last few years it has been found to be central to heart disease, cancer, diabetes, stroke, arthritis and neurodegeneration, i.e. all the diseases of aging. The speed with which researchers are jumping on the inflammation bandwagon is breathtaking. Just a few years ago, 'nobody was interested in this stuff', says Dr Paul Ridker, a cardiologist at Brigham and Women's Hospital in the US, who has done some of the groundbreaking work in the area. 'Now the whole field of inflammation research is about to explode.'

It was Ridker who discovered a link between heart disease and inflammation. In 1997 he found that patients with high levels of chronic inflammation were three times as likely to suffer a heart attack, whereas those with low levels of inflammation were unlikely to suffer an attack. How does this work? A heart attack is usually caused by a blood clot getting stuck in, and thus blocking, the blood vessels supplying the heart. Hence the heart muscle gets no oxygen and stops working, so the heart stops pumping blood around the body. The blood clot usually arises from rupture of a fatty (arteriosclerotic) plaque on the wall of the heart (coronary) arteries. But why is it that in some people these fatty plaques rupture, whereas in other they are stable and never break open to generate a blood clot? Ridker believes that it is inflammation that causes the plaques to rupture. Ridker says. 'Cholesterol deposits, high blood pressure, smoking – all contribute to the development of underlying plaques. What inflammation seems to contribute is the propensity of those plaques to rupture and cause a heart attack'.

Inflammation may be important to the formation of the plaques too. Cholesterol, high blood pressure and smoking can all damage the blood vessel wall. The damage triggers inflammation in the wall, which in turn attracts white blood cells into the wall that generate the fatty plaques. So blocking the inflammation with aspirin or statins may protect against heart attack and stroke in several ways.

Diabetes is characterized by an uncontrolled blood sugar level. It comes in two flavours: juvenile-onset or adult-onset, depending on the most common time at which disease starts. It has long been known that juvenile-onset diabetes is due to the immune system destroying the cells within the pancreas that generate insulin, the hormone that regulates blood glucose level. What is new is that adult-onset diabetes may have an inflammatory component. This type of diabetes is often caused by obesity, and it was recently discovered that fat tissue generates a particular type of hormone (known as cytokines) that regulate inflammation. And these cytokines from the fat cause the rest of the body to become insensitive to insulin, and this in turn causes the diabetes.

Back in the 1860s, the famous pathologist Rudolf Virchow speculated that cancerous tumours arise at the site of chronic inflammation. Recent evidence suggest that he may have been right. Cancer is caused by mutations of the cell's DNA, but what is causing the mutations? A lot of people are starting to think that the mutations are caused by the free radicals released during inflammation. Indeed it has been suggested that there is a natural progression from acute inflammation, to chronic inflammation, to cancer. Cancer may be a consequence of any chronic inflammation that is not resolved. A cell that has sustained a few mutations may be dysfunctional, but it is rarely cancerous. On the other hand the immune system may detect that the cell is not normal and trigger inflammation, but instead of destroying the abnormal cell the inflammation may generate more free radicals that produce more mutations that eventually turn the cells cancerous.

To the immune system, a developing cancer may look very much like a wound that needs to be fixed. Lisa Coussens of the University of California, San Francisco, explains:

When immune cells get called in, they bring growth factors and a whole slew of proteins that call other inflammatory cells. Those things come in and go 'heal, heal, heal.' But instead of healing, you're 'feeding, feeding, feeding'.

What starts this inflammatory cycle that generates cancer? In lung cancer it may be the cigarette smoke that damages the lung lining, which then triggers chronic inflammation that results in lung cancer. In the gut, it may be stomach acids or components of the food that damage the gut lining to trigger inflammation that eventually results in colon cancer. Taking aspirin or other anti-inflammatory drugs has been shown to prevent the development of pre-cancerous growths in the gut known as polyps.

The humble aspirin is turning out to be one of medicine's most powerful tools – we just didn't know it. Medics stumbled on the startling fact that patients taking aspirin or other anti-inflammatory drugs for their arthritis or heart disease were also strongly protected against developing Alzheimer's disease. Your chances of developing Alzheimer's can be reduced by up to 50% by taking such drugs continuously and long term, i.e. for 10 years. Subsequently scientists have found that the brains of Alzheimer's patients are strongly inflamed, probably because the β-amyloid protein (which is the hallmark of this disease) itself may provoke inflammation. And free radicals produced during inflammation may kill neurons, resulting in dementia.

Several other neurodegenerative diseases have an inflammatory component. For example, Parkinson's disease patients appear to lose neurons due to inflammation deep inside the brain, and taking anti-inflammatory drugs long term seems to protect against the development of the disease. Multiple sclerosis results from the attack of the immune system on neurons and support cells in the brain and spinal cord. It can be controlled by particular types of anti-inflammatory drug, but not aspirin. Many other diseases caused by viruses (e.g. AIDS), bacteria (e.g. meningitis), or protozoa (e.g. malaria) cause brain damage by provoking inflammation in the brain, which kills the pathogens but may also wreck the brain.

In my laboratory in Cambridge, we investigate how inflammation causes damage and disease, in particular neurodegenerative disease. We have found that most of the damage is
inflicted by a particular type of brain cell, known as 'microglia'. Microglia are small and few (about one in ten cells in the
brain are microglia), and normally very quiet. The resting
microglia are however restlessly vigilant. They rove around the
brain and put out long feelers everywhere to search for
damage or pathogens. As long as there are no pathogens or
damage, the microglia remain quiet. But when they detect
something wrong, using their pathogen and damage-detec-
tion systems, they initiate a frenzy of activity. They move
swiftly to the detected danger area, where they rapidly prolif-
erate and release large quantities of danger signals that attract
other microglia and white blood cells, and alert surrounding
cells to the danger. They also release massive doses of free radicals and other toxins in an attempt to sterilize the whole area.
They start to engulf and eat anything in the vicinity. Then they
digest what they have eaten and present it to white blood
cells, which may then engage the adaptive immune system to
attack the area. The microglia are involved in a desperate
battle to stop damage and disease from spreading in the
brain, and they may not be too worried if a few neurons get
killed in the crossfire.

Normally the microglia are very effective in their job: they
clear the pathogens and damage quickly and return to a quiet
resting state. However, some types of damage and pathogen
are resistant to the attack of the microglia, so the microglia are
chronically at war without ever winning. Just as a country can
be devastated by a chronic war, so the chronically infected or
neurodegenerating brain is devastated by an unresolved war
between microglia and pathogens or damaged proteins. In
Alzheimer's brains, the β-amyloid activates the microglia, so
the microglia try to eat the β-amyloid, but they can't digest it,

so they end up being chronically activated by it. In Parkinson's disease the microglia are inflamed by the synuclein protein that makes up Lewy bodies. In AIDS, the virus climbs right inside the microglia and disables its protective mechanisms, causing inflammation of the microglia but without being able to destroy the virus.

Inflammation contributes to other diseases of aging, such as arthritis and osteoporosis, and possibly to the non-disease aspects of aging. The aged body and brain suffers from a 'smouldering inflammation': chronic, low-level inflammation, possibly caused by accumulating damage, damaged proteins or pathogens. Inflammation almost always disturbs the function of the tissue it affects, but chronic inflammation can completely block those functions and lead to the diseases we have been talking about. However, there is no evidence that blocking inflammation can block or slow aging, so it probably contributes little to the aging process itself, in contrast to the diseases of aging.

Inflammation protects us from acute damage and disease, but the free radicals and other toxins it produces may in the long term cause the degenerative diseases of aging. This is exactly what we would expect from the antagonistic pleiotropy theory of aging that we discussed at the beginning of this chapter. Inflammation protects us from acute causes of death (wounds and infection) but as an unintended side-effect, it also causes damage to our body that accumulates with age. Because we evolved in conditions when death from infectious disease and wounding was ubiquitous, whereas deaths from old age and diseases of old age were rare, our bodies and immune systems have evolved to fight infection hard without caring about the collateral damage that eventually results in the diseases of middle and old age. The chronic death that currently afflicts our society is partly a legacy of our ancestors battling acute death during thousands and millions of years of evolution by natural selection.

Is there a realm of youth beyond extreme old age?

Recently biologists have come across startling evidence that aging may slow and even reverse in extreme old age[16,17]. This has led some to suggest that there is a third phase of life beyond development and aging: immortality. Others have denounced these ideas as crackpot.

Michael Rose of the University of California at Irvine used fruit flies to investigate how susceptibility to death depends on age[16]. Young flies rarely died in the lab, but the number dying per day increased expontentially with age, as predicted by the Gompertz relation. However, at extreme old age (which for flies is only a month) susceptibility to death plateaued off, as if aging no longer increased beyond a certain age. This strange observation has been repeated in a variety of other short-lived animals, and in some susceptibility to death appears to decrease in extreme old age[17]. Close examination of the rate of dying of humans also suggests that it plateaus beyond the age of 100 years, and may even start to decline beyond 110 years.

Perhaps aging is like crossing the Himalayas. As we journey upwards into the thin air of late old age our susceptibility to death increases dramatically the higher we go. But a few hardy souls make it over the top to a high plateau, where life is difficult but not impossible. Is this the Shangri La of immortality?

The truth is that so few individuals have made it to extreme old age that we really don't know. These survivors are unlikely to be representative of the general population. They must be special in some way (genetically, culturally, or environmentally) in order to have survived so long. And therefore the rate of dying of these extreme survivors may be lower than that of the general population, leading to a spurious, apparent plateau in the rate of dying with age. The same criticism can be made of the animal studies – those that survive to extreme old age may be unrepresentative of the general population.

In the twenty-first century we are going to be sending not dozens, but millions of people up into the thin air to explore the mysterious realm of life beyond 100 years of age. These explorers will eventually discover whether aging and death continue to rise with age, or whether they slow down and perchance reverse. Perhaps a few hardy explorers will reach that Shangri La of reversed aging, and so will have uncovered a new realm of life. Only time will tell.

interlude 8
should death be resisted?

Should death and aging be resisted? Dylan Thomas appeared to think so, as he recommended: 'Do not go gentle into that good night, Old age shall burn and rave at close of day; Rage, rage against the dying of the light'. He calls on us to reject death and the forces of darkness 'though wise men at their end know dark is right'.

However, surprisingly throughout history we have been told that death is natural and not to be resisted. The Preacher of *Ecclesiastes* asserts: 'To every thing there is a season, and a time to every purpose under the heaven: A time to be born, and a time to die'. Before that Homer had written, 'The race of men is like the race of leaves. As one generation flourishes, another decays'. Near the end of his life the third US President, Thomas Jefferson, wrote to the second President, John Adams: 'There is a ripeness of time for death, regarding others as well as ourselves, when it is reasonable we should drop off, and make room for another growth. When we have lived our generation out, we should not wish to encroach on another'. Even evolutionary biologists, such as August Weismann and Alfred Wallace, asserted that aging and death were evolutionarily designed by natural selection to make room for the next generation and thus benefit the species.

There have been occasional dissenters who have violently resisted death. One of the most celebrated is the Greek antihero Sisyphus. He escaped death twice, he chained Death to a rock, and he tricked his way out of the underworld. For this the gods condemned him to ceaselessly rolling a huge rock up a hill; it would then roll back down and he would start again... and again. Life can seem absurd, and our own lives are analogous to the task of Sisyphus. For obscure reasons, as if ordained by the gods, we toil away throughout life, creating a self, a family, a career, possessions, pride, only to have everything snatched away again by aging and death. Sisyphus could do little about his fate, but that did not stop him spitting in the face of death, and scorning the gods and the fate they had condemned him to.

In recent times, partly as a consequence of the rise in science and decline in religious belief, an 'Immortalist' movement has developed, particularly in the USA, which is eloquently described in Bryan Appleyard's book How to Live Forever or Die Trying[1]. These Immortalists believe that we should move heaven and earth to bring about human immortality now. Alcor and the Cryonics Institute are two US companies that advocate and practice freezing people immediately after death in the hope that they may be revived in some future generation. The Extropy Institute and the World Transhumanist Association advocate scientific enhancement of humans (including immortality) to the point that we are no longer human. The Singularity Institute in California purses Artificial Intelligence routes to immortality. The American Academy of Anti-Aging Medicine pushes mainstream medicine to increase longevity. The Methuselah Foundation runs the Mprize to encourage research into longevity. The Buck Institute in California houses and funds scientists investigating aging and extending longevity. SENS (Strategies for Engineered Senescence) is a loose alliance of scientists lead by Cambridge Immortalist Aubrey de Grey that pursues strategies to engineer immortality in the human body. The Immortality Institute (Imminst) was founded in 2002 with the aim of: 'conquering the blight of involuntary death'. Mortality is under attack, and the attack is coming from the USA.

Why are all these different institutions seeking immortality? As Imminst's founder Bruce Klein says: 'oblivion is the issue'. Everyday 150,000 people die in the world, 100,000 of them from diseases of old-age. Klein believes this is a disaster, to which we should respond as to other disasters:

I call it the silent Tsunami, every day more than 100,000 people die quietly and acceptingly, saying their time has come, or some other such euphemism. But, with a real tsunami, they say this is a tragedy, we must do things to prevent this in the future.

With science suggesting that mortality is no longer inevitable, some Immortalists feel that we are now morally bound to at least try to prevent the slaughter by seeking an end to aging and death.

One of the emblematic leaders of the Immortalist movement is American Ray Kurzweil, inventor of various software systems enabling computers to read, recognize speech, play the piano, play the stock market etc. He believes that the exponential growth in computer power and artificial intelligence (AI) will soon enable computers to design and construct themselves for further levels of power and AI. This will lead to runaway growth of AI and eventually the 'singularity', where AI computers far exceed the capacities, intelligence and even comprehension of humans. Kurzweil believes that the singularity will give us the AI power to engineer immortality into our bodies. But what should we do while we are waiting for the singularity?

In Kurzweil and Terry Grossman's book *Fantastic Voyage: Live Long Enough to Live Forever.* They advocate a three-step path to immortality. First, do everything currently known to extend your life. This includes living healthily, optimum diet, exercise, taking supplements, and getting regular extensive health checks. That should help you live long enough for step two to come into operation, which is the application of current biotechnologies, such as gene therapies and stem cell research. And this in turn should help you live long enough for step three to have been developed, which is nanotechnological repair and reengineering of the body.

Kurzweil is currently applying step one to himself, including taking 250 supplements a day and spending a whole day a week in clinic:

Whereas some of my contemporaries may be satisfied to embrace aging gracefully as part of the life cycle, that is not my view. It may be 'natural', but I don't see anything positive in losing my mental agility, sensory acuity, physical limberness, sexual desire or any other human ability. I view disease and death at any age as a calamity, as problems to be overcome. Until recently there was relatively little that could be done about our short lifespan other than to rationalize this tragedy as actually a good thing.

Kurzweil is the caricature of an American baby-boomer, extravagantly individualistic, optimistic and ambitious. In *Fantastic Voyage* he notes the perilous position of his generation, poised on the edge of the 'critical threshold' of life-extension technologies:

A small minority of older boomers will make it past this impending critical threshold. You can be among them.... Unfortunately most of our fellow baby boomers remain oblivious to the hidden degenerative processes within their bodies and will die unnecessarily young.

Transhumanists believe that we should seek to artificially enhance humans to the point that we are no longer humans, but rather something better than humans, and that includes living longer or 'forever'. For example, Kurzweil suggests that the fusion of AI, molecular biology and nanotechnology will create molecular robots on the nanometre scale, nanobots, which when incorporated into the brain will enable a hybrid of both biological and non-biological intelligence:

This will enable you to profoundly expand your pattern recognition, and emotional capacities as well as provide intimate connection to powerful forms of non-biological thinking. You will also have the means for direct high bandwidth communication to the Internet and other people from your brain.

Nick Bostrom is director of the Oxford Future of Humanity Institute and co-founder of the World Transhumanist Association. He believes that all previous utopian, political movements have been broken on the hard rock of human nature, so the only way forward is to change that human nature by technological means. Bostrom wrote what has become almost a sacred text amongst Immortalists: 'The Fable of the Dragon Tyrant'. This is a story of an evil dragon that rules over a helpless people. It demands a blood tribute of ten thousand men and women every evening. Sometimes it devours them immediately, sometimes it locks them up in its lair to wither away for years before it finally consumes them. The dragon is, of course, death. Religions arise to justify the slaughter or to console the people with promises of life-after-consumption. But a prophet appears to predict that one day the people will slay the dragon. And indeed a missile is devised to blast the dragon to kingdom-come. There is great resistance to this plan from the conservative and religious types: 'The phrases were so eloquent that it was hard to resist the feeling that some deep thoughts must lurk behind them, although nobody could quite grasp what they were'. Finally the deed is done – the dragon is slain – and at first the people are distraught that it was not done years ago, saving millions of lives – but then they embrace their immortal and transhuman future.

Bioconservatives, such as Francis Fukuyama, have argued that immortality is not desirable precisely because it would make us transhuman. This echoes an ancient Greek idea that humans are the only mortals. The gods and nature are immortal, while animals are not individuals, and therefore cannot die. Mortal humans are in contrast individuals, who know they are to die. Fukuyama argues that mortality is essential to being human – all our most cherished beliefs, motivations and loves are based on mortality. Without mortality we could not want, or care, or love, or strive, and we would be something entirely different, that had lost something essential to being human.

Boredom is another potential danger of immortality. In *Thus Spoke Zarathustra*, Nietzsche asks the question 'How well disposed would a person have to become to himself and to life to crave nothing more fervently than the infinite repetition, without alteration, of each and every moment?'. Many people become tired with themselves and the world after just one lifetime. Nietzsche believed that most people, becoming immortal, would eventually find the infinite repetition of themselves, the world and all its quirks as a kind of Hell. The person who could accept recurrence without self-deception or evasion would be a superhuman being (Übermensch), a superman whose distance from the ordinary man is greater than the distance between man and ape. In a sense then, the superhuman who could bear to live forever would be a transhuman or god. Immortalists have claimed that boredom could be rectified either by natural memory loss (so we only fully remembered the last few decades) or by reengineering the brain (so we change the self or have a limitless capacity for wonder). 'Deathists' have countered that such remedies amount to death of self, so why bother avoiding natural death?

Immortalists have been inspired by the revolutionary conclusion of twentieth century biologists (outlined in the last chapter) that aging is not a natural or designed process, and serves no biological or evolutionary purpose. Previous generations of Deathists, by contrast, had assumed that aging was natural, was programmed into our bodies, and served some higher purpose. However, modern-day Deathists have claimed that immortality is not generally desirable for the individual and/or society because of overpopulation, reduced resources for the young, reduced turnover of ideas, and inequity between mortals and immortals. Alternatively, they argue that it is unnatural, interferes with God's business, or messes up intergenerational relations. Most of these objections could be applied to any other medical intervention to extend life, yet we normally think of extending life as being a good thing. We would be outraged if it was suggested that someone's cancer should not be treated because they were old, but we might think differently if all causes of their mortality could be cured. So what's different about immortality?

Many people have a creeping feeling that there is something selfish or sacrilegious about wanting immortality. Certainly the God of Genesis was not keen on the idea.

And the Lord God said: 'The man has now become like one of us, knowing good and evil. He must not be allowed to reach out his hand and take also from the Tree of Life and eat, and live for ever'. So the Lord God banished him from the Garden of Eden to work the ground from which be had been taken. After he drove the man out, be placed on the East side of the Garden of Eden cherubim and a flaming sword flashing back and forth to guard the way to the Tree of Life. (Genesis 3: 22–24)

This God appears to have believed that the only two things separating himself from man were knowledge and immortality. Man had already acquired knowledge, separating him from the animals and the state of nature, so it was essential that he did not acquire immortality from the Tree of Life, and thus become a god.

Of course, this is not the current Christian interpretation of Genesis. The predominant Catholic interpretation, derived partly from thiteenth century mystic St Thomas Aquinas, would have it that in Eden man was united with God in a state of grace, which included Immortality. It was because man defied God that they were separated, so that man fell from grace into sin and mortality. But through the death of Jesus Christ and death of our body, we may regain that state of grace and immortality in a spiritual form. Christianity put death (and suffering) at the centre of religion: only through death could we truly live. Therefore death should not be resisted, but rather embraced.

However, according to Nietzsche, God is now dead. So can we creep around the flaming sword and cherubim, back into the Garden of Eden, steal the fruit of Immortality, and become gods ourselves? It seems like a tall order, but as we shall see, perhaps not impossible.

9
immortality

In this final chapter we need to bring together the rest of the book to form some conclusions about the future of death. But we can also go beyond that to consider the prospects for achieving immortality in the twenty-first century, and if this is possible, on what terms.

Is there a maximum human lifespan?

Average life expectancy has increased by about 2–2.5 years every 10 years for at least the last 100 years (Figure 2.1). The future of humanity depends on whether this rate of increase in lifespan itself increases, decreases or stays the same. All three options have dramatic implications, both for ourselves as individuals and for the human species. If the rate of increase stays the same over the next 100 years, then an extra 25 years will be added to life expectancy, and the world will be dramatically aged, with all the health and economic consequences that implies. If the rate decreases towards zero, then old age itself may disappear from human experience. If the rate increases to 10 years per 10 years, then, as pointed by futurologists, the consequence is effectively immortality. But at what cost?

There has been considerable controversy amongst demographers over the last few years as to which of these fates is most likely. Jay Olshansky at the University of Illinois and Bruce Carnes at the University of Chicago have pushed the idea that longevity increases are bound to slow down, due to: the difficulties of further reducing chronic disease[1], the obesity epidemic[2] and biological limits to the body[3]. In 1990 they rashly predicted that for the foreseeable future 'it seems highly unlikely that life expectancy at birth will exceed the age of 85'. By 2005 life expectancy for women in Japan was 85.5 years, and was still increasing at two years per decade.

In contrast, James Vaupel of the Max Planck Institute for Demographic Research, Rostock, Germany, has pointed out how rising life expectancy has continued to confound demographers by breaking all their previous predictions of limits to life expectancy[4]. He sees no sign of rising life expectancy slowing down in the near future. Indeed rising life expectancy amongst the oldest fraction of the population has accelerated in the last few decades: Death rates halved for women in their 80s between 1950 and 1995 in developed countries. And aging may decelerate or even reverse in extremely old people[5].

Although the *average* human lifespan has been increasing steadily for the last hundred years or so, it is unclear whether the *maximum* lifespan has changed at all. The concept of maximum lifespan is controversial and possibly meaningless. The maximum could be the oldest age anyone lives to or has ever lived to. But it is difficult to know this for past ages because records are poor, and anyway it refers to a single exceptional individual that may be susceptible to randomly varying factors. Alternatively, maximum lifespan can be calculated as the age at death of the longest-lived 1% of the population, or extrapolated from plots of population survival against age.

The idea that the maximum lifespan of humans is 120 years is often mistakenly presented as scientific fact, but actually comes from higher authority. Just before the biblical flood, God apparently said: 'My Spirit will not remain in man forever, for he is mortal; his days will be a hundred and twenty years' (Genesis 6:3). Prior to the flood, the Old Testament tells of an era when maximum lifespan approached 1000 years. Adam lived 930 years and Noah 950 years, while Methuselah holds the record at 969 years. Hence Methuselah is the individual with the oldest 'recorded' age. Coincidentally it was recently estimated by gerontologist Stephen Austad that in the absence of aging the human lifespan might be 1200 years (if there were no aging, but death still occurred from extrinsic causes such as accidents).

The biblical story of Methuselah *et al.* was the inspiration for many Antediluvian legends – legends of a Golden Age in the distant past when people lived much longer. There is no archaeological evidence for such a past – no bones suggesting unusual age, and no credible historical records of excessive age. So the Antediluvian legends are probably just that: legends. The oldest authenticated age to which any human has ever lived is 122 years and 164 days by Jeanne Calment, a French woman, who died in a nursing home in 1997. So we know for a fact that our maximum lifespan (if such a thing exists) is greater than 120 years, and the greatest was probably achieved relatively recently, suggesting that 'maximum' lifespan is increasing. However, credible records of age at death only exist for the relatively recent past, so it is impossible to say definitely whether maximum lifespan is greater now than in the far past.

If there were a fixed maximum lifespan, we might expect the increase in average life expectancy to slow down as it approaches this theoretical maximum. Demographers have been predicting such a slowing for decades. The trouble is that experience has always shown them to be wrong, as actual life expectancy has continually ruptured these supposed barriers. If human life expectancy was approaching a maximum we should see a decline in the rate of increase in those countries with the highest life expectancy as it approaches the maximum. Jim Oeppen and James Vaupel investigated this very question, and found that if they plotted life expectancy against date in the country with the highest expectancy in the world (Sweden, Norway, New Zealand and now Japan) the rate of increase has been remarkably constant (at 3 months per year) all the way back to 1840, and shows absolutely no signs of slowing down (Figure 2.1).

Japan has the longest life expectancy: 85 years in women, 78 in men in 2002, and there is no sign of the rate of increase dropping off. The number of Japanese citizens aged at least 100 years has doubled over the last five years, numbering just over 20,000 in 2004. Centenarians are no longer rare exotic species: they are the fastest growing section of the population. By 2030, the over 65s are expected to account for almost 30% of the population. If current trends continue female life expectancy in Japan will be 97 years by 2050, and average (male plus female) life expectancy 100 years by 2065. This is the future: and it's old.

Leading aging researcher Tom Kirkwood has commented:

> Demographic trends over the past 2–3 decades have shown that, instead of reaching a plateau as most forecasters had predicted, human life expectancy has continued to increase and shows no sign of slowing. This has taken everyone by surprise, because the conventional view of aging is that maximum life expectancy is somehow 'fixed'. Thus it was expected that when preventable early deaths were brought under control by medicine, we would simply see the aging process more clearly revealed, with more and more of us living lifespans approaching the intrinsic limit. The rapid declines in old-age mortality over the past 50 years show that this is not happening, which has led us to re-examine the very nature of the aging process.

Although the average human lifespan has been increasing steadily for the last two hundred years or so, the rate at which we age appears to have changed little or none[6]. The average lifespan has increased because we have removed specific causes of death, but the underlying molecular processes of aging within the body have continued unabated. Because average life expectancy is increasing, but the rate of aging is unchanged, years spent in ill health and dependency are increasing, and people are increasingly dying of 'old age'[7]. It has been calculated that if we got rid of the three main causes of death today (cardiovascular disease, cancer and stroke) then life expectancy would only be increased by 15 years[7,8]. Fifteen years is not bad, but it's not immortality either. And those extra 15 years would not be added on to youth, but rather spent in (very) old age.

We are not even sure what old people do die of in the absence of cardiovascular disease, cancer and stroke. Autopsies are rarely done in such circumstances. The death certificate will record some immediate proximal cause, such as pneumonia, but this is just the executioner of death, not the mastermind behind the death. The ultimate cause is aging, but in between the ultimate cause and proximal cause there is likely to be some specific disease process or body dysfunction that led to a specific form of death. In the case of pneumonia ('the old man's friend') the reason the old man becomes susceptible to infection by the pneumonia bacteria, when a few years earlier he would have easily defeated such infection, must result from dysfunction in the lungs or immune system or other body systems controlling these. As people live longer, and fewer people die from known causes, we will inevitably discover new disease processes associated with advanced aging, just as advancing longevity has already revealed the neurodegenerative diseases.

The vast majority of aging research is done on age-related diseases, such as cancer and cardiovascular disease, rather than the aging process itself. This may be counterproductive because, as already pointed out, this may extend life a little, but the quality of extended life will be poor. Only by tackling the aging process itself will we have any chance of substantially extending life quality as well as life quantity.[7]

On the other hand, aging may not be one single, all-power-ful, active killing process. It may rather be a large number of passive failures to prevent dysfunction that manifest them-selves as diseases of different systems at different times, in which case aging may merge into diseases of aging. Most death nowadays is ultimately caused by aging. But, by giving aging a name, we are in danger of reifying it to the status of a single entity or force, and then trying to attack this imaginary entity. We do not think of disease as a single entity to be attacked; rather we distinguish different disease processes than can be studied and treated. Similarly, perhaps we should not think of aging as a single process but rather distinguish a set of different aging processes that can be studied and treated separately.

We cannot be absolutely sure whether the rate of aging is increasing, decreasing or staying the same. What we can say is that over the last 150 years, when average lifespan has dou-bled, there has been no obvious change in the rate of aging. Photographs, testaments, novels and medical records indicate that the average 65-year-old in the nineteenth century looked and functioned in pretty much the same way as the average 65-year-old today[12]. There are a variety of differences of course: for example, the rate of mortality is lower today, and health is on the whole better, there are cosmetic means of hiding aging, and people are fatter today (but as noted in the previous chapter the latter has been suggested to speed up aging). So it depends how you define aging, but the intrinsic rate at which our cells senesce appears to have changed very little over this time-scale. Further back in time, literature, art, historical records, and forensic archaeology all indicate that the rate of aging has changed little during recorded his-tory[9,10]. For example, the mortality doubling time, a measure of aging rate, is thought to have been the same, even back to the Stone Age[10].

Immortality is not the same thing as eternal youth. This was the disastrous confusion that Tithonus made. Although life expectancy has risen and continues to rise rapidly, aging and the rate of aging appear to have remained almost unchanged. Because aging appears to increase exponentially with age, the extension of life may produce some spectacularly aged individuals. However, all this may be prevented by the 'compression of morbidity'.

Compression of morbidity

If there were a maximum human lifespan, the gradual increase in the average lifespan towards this figure, together with the medical delaying of age-related diseases, would result in people dying closer and closer to the maximum, with less and less age-related disability and disease. This optimistic concept is known as the 'compression of morbidity', and was developed in 1980 by Dr James Fries, a physician from Stanford University. Fries postulated that 85 years was the biological limit to human lifespan, and he predicted that in the future death will become increasing concentrated around this age (the compression of mortality). And he further predicted that improved lifestyles and medical advances would concentrate the age-related diseases into the same narrow age range (compression of morbidity). Mortality refers to the probability of death, and morbidity refers to the probability of disease. So Fries was predicting that in the future we would approach the ideal of all living full, disease-free lives up to the age of 85 and then promptly drop dead.

This optimistic prediction was immediately controversial and considered to be naïve, because there was and is no convincing evidence for an absolute maximum lifespan or the ability to significantly delay the age of onset of degenerative diseases. Critics of the compression theory pointed out that without the development of more effective treatments for the non-fatal diseases of aging, the increasing average lifespan due to preventing fatal diseases would inevitably result in a longer period of disability and disease prior to death. This pessimistic view of our future is known as the 'expansion of morbidity' theory, and was developed by Ernest Gruenberg and others in the 1970s, and has been elaborated by many other scientists since. They also believe that death rates from killer diseases will continue to be reduced. But in contrast to the 'compression of morbidity' theorists, they believe that there is no fixed upper limit to lifespan and that the age of onset of non-fatal disease is unlikely to be significantly delayed. The result is Tithonus's syndrome: extended old age with expanded disability, disease and dementia.

These two competing visions of the future (expanding youth or expanding old age) cannot both be right. Which is correct has huge consequences for society. If we just think of the economic consequences: An endlessly expanding population of retired, seriously ill old people, living longer and longer, and requiring more and costlier health care, predicts economic stagnation or collapse. Thus many gerontologists (medics or scientists who study old age and aging) and demographers (researchers of trends in population statistics) have tried to determine which vision of the future is closer to the truth. However, it has turned out to be very difficult to get an unambiguous answer to this question, as different researchers with different interests and agendas used different statistics to construct their answer.

In 1984 a group of gerontologists and demographers suggested to the World Health Organization (WHO) a more objective means of getting at the answer. They proposed a general model of aging and how to quantify it using population statistics that distinguished between total survival, survival without disability, and survival without chronic disease. So they proposed calculating, for each population, the average number of years of life (life expectancy), the average numbers of years of life without disability (disability-free life expectancy) and the average number of years of life without chronic disease (disease-free life expectancy).

Following this model, an international organization of demographers (REVES) was set up in the 1980s, headed by the Frenchman Jean-Marie Robine. They came up with the term 'health expectancy' to quantify the years of life spent in good health, rather than ill health. So life expectancy could be divided into health expectancy and ill health expectancy, and the demographers set out to determine which was increasing. Research in the 1980s and 1990s indicated that life expectancy was increasing faster than disability-free life expectancy in most parts of the world[11], supporting the 'expansion of morbidity' theory, i.e. the Tithonus scenario.

However, research in the 1990s suggested the trend was tilting back towards compression of the years of ill health in the US. More generally it was found that years of moderate ill health and disability were increasing, while years of severe ill health and disability were decreasing (a model of population aging known as 'dynamic equilibrium').

This research resulted in a burst of optimism among gerontologists, and coincided with popularization of the concept of 'successful aging', an idea originating in 1987 from John Rowe and Robert Kahn in the US. 'Successful aging' became a research programme aimed at identifying characteristics shared by those people who survived old age with intact mental and physical health. In particular they wanted to discover why some people can continue to perform daily activities such as walking, toileting and dressing into extreme old age (successful aging), while others cannot. The laudable aim was to spread best practice from the successful agers to the unsuccessful agers. However, it remains unclear whether successful aging can be significantly controlled by the aged person, rather than by genetics or environmental factors outside their control. In these circumstances the successful aging concept can end up designating most old people as failures, without them being able to do anything about it.

Research in the 2000s indicates that the optimistic trends seen in the 1990s may have been just a blip in a longer-term trend towards increased ill health expectancy. Expected years of ill health are currently rising at the rate of approximately one year per decade in the UK (Figure 4.4), while expected years of health have more or less stopped increasing (Figure 4.4). Similar long-term trends have been seen in many countries[11]. Thus morbidity is currently expanding. This has massive consequences for society in terms of economic growth, as well as personal consequences in terms of the quality of life.

Is the Tithonus scenario inevitable?

A major determinant of whether morbidity at the end of life will expand or compress is dementia. The future growth of dementia is of huge significance to society, partly because of the frightening costs in time and money of looking after a demented patient, and partly because of the frightening quantity of people involved. Just how many people is currently a matter of contention. The proportion of people with dementia appears to double every extra five years of age, so that of those aged 65 years in the US, 2–3% show signs of Alzheimer's, at 75 years of age roughly 10% have Alzheimer's, and at 85 years of age the probability of having Alzheimer's is 25–50%. If this trend were to continue then those poor souls reaching 100 years or beyond would have a near certainty of developing Alzheimer's.

However, in fact only 25–50% of those reaching 100 have signs of Alzheimer's. What this means is unclear – partly because the current centenarian population is small and highly selected: centenarians must be much more resistant to age-related disease than the general population in order to have survived so long, and thus are unrepresentative. The proportion of the population most susceptible to Alzheimer's may have been killed off before reaching 100 years of age by the disease itself, which takes about 10 years to kill. However, other neurologists think that the probability of getting Alzheimer's declines above the age of 85. And as we have seen, some believe that aging itself slows in very old age.

Hence there is some uncertainty when projecting the prevalence of Alzheimer's into the future, when the numbers of very old and extremely old people are predicted to increase dramatically. Further uncertainty is caused by the diagnosis of Alzheimer's or dementia, which differs in different countries. The disease is often undiagnosed or unreported, and can be difficult to distinguish from normal brain aging when someone is extremely old. But the greatest uncertainty is in predicting the numbers of very old and extremely old people in the future: does one use the *current* probabilities of dying at different ages to predict the future population, or does one use current *trends* in probabilities of dying to predict future probabilities and populations?

Despite these difficulties, it is important to try. The most recent (2003) US study of Alzheimer's prevalence, using the 2000 US census, indicates that there were 4.5 million US Alzheimer's sufferers at that time, and this level was predicted to increase by 300% by 2050 (Figure 2.5). Using low, middle and high US Census projections, the study predicts that by mid-century, between 11 and 16 million Americans will have Alzheimer's. According to Sheldon Goldberg, head of the US Alzheimer's Association, 'If left unchecked, it is no exaggeration to say that Alzheimer's disease will destroy the health care system and bankrupt Medicare and Medicaid'.

European predictions appear to differ from their US equivalents, partly because European estimates have been based on diagnosed Alzheimer's disease, and this may underestimate the true prevalence because many people with Alzheimer's are not diagnosed as such. But the incidence of Alzheimer's may be genuinely higher in the US, where 46% of people over 85 years of age are thought to have the disease. In the UK, recent estimates indicate that there are currently about 700,000 people (1% of the population) with dementia (two-thirds of which is due to Alzheimer's), and there will be 1.7 million by 2050[12]. That is a lot of people, but it still does not take into account recent updates in the predicted number of very old people in the UK. These predictions are continually being increased because they are generally based on current probabilities of dying (i.e. current longevity), and those probabilities keep decreasing (i.e. longevity keeps increasing).

If we extrapolate current trends in life expectancy in the UK (Figure 4.3), then we would predict a life expectancy of 91 for women and 87 for men in the year 2050. Recent estimates indicate that the current prevalence of dementia in the European Union and UK of people between 90–100 years is about 33%. Current US estimates of the probability of having Alzheimer's at 85 years are 25–50%. Thus, if the probability of having dementia at any given age does not change, then by 2050 about one third of people dying would be dying with dementia.

In recent years a precursor to Alzheimer's disease has been identified, known as Mild Cognitive Impairment (MCI). Symptoms include memory decline (including forgetting names and objects), not recognizing family and friends, forgetting one's own phone number or address, difficulty finding a familiar place, and/or noticeable language and intellectual decline.

It is still unclear how common this condition is, but a recent estimate was 1% of 60-year-olds rising to 42% of 85-year-olds, while Alzheimer's was assumed to affect 1% at 60 years and 25% at 85 years[13]. If this is true, two out of three 85-year-olds has either MCI or Alzheimer's. This is not encouraging, even on current life expectancy figures: in 2005 life expectancy for 65-years-olds in the UK was almost 17 years for men and 20 for women. Thus roughly two thirds of women who are 65 years old now would be expected to develop either Alzheimer's or MCI before they died.

The feasibility of the Tithonus scenario coming to pass depends partly on how quickly dementia kills the sufferer. If dying with Alzheimer's becomes common because dying from other causes of death is reduced, then Alzheimer's will be a major cause of death, and further increases in longevity will be impossible without curing Alzheimer's. Alzheimer's currently kills patients on average 8–10 years after diagnosis, although many patients die before then from other causes because they are old or very old. Removing other causes of death not only increases the likelihood of living long enough to get dementia but also the number of years lived with dementia, and thus its potential severity.

Future medical advances may impact on Alzheimer's in a number of different ways. An intervention (such as a vaccine) might prevent or delay the onset of the disease – this would eliminate or compress the number of years lived with the disease. Alternatively, an intervention (such as a drug) might prevent the disease from progressing (or slow its rate of progress) once it has started. This would potentially expand the number of years (and the number of people) living with the disease, although reducing its severity. While the former type of intervention is more desirable, past medical and pharmaceutical history would suggest that the latter intervention is much more likely.

Is it really feasible that life expectancy will continue to increase at the rate 2–2.5 years per decade, so that by the end of century it could be 100 years? Most people nowadays die from the age-related degenerative diseases: cancer, cardiovascular disease and stroke. As we have seen, most of these diseases are on the decline as a cause of death (Figure 2.3) or people are living longer and longer with them. A partial exception is cancer, where some cancers are increasing (mainly due to increasing life expectancy) and some are decreasing. Future reductions in smoking could substantially reduce deaths due to cancer, cardiovascular disease and respiratory disease; and changes in diet and air quality could further reduce deaths.

On the other hand, the current obesity epidemic is tending to push up cardiovascular, cancer and diabetic deaths. However, given the amount of basic, medical, epidemiological, genetic and pharmacological research currently devoted to these diseases, and the current rates of progress in these areas, it is conceivable that these might be 'cured' sometime this century, at least in the sense of being converted to chronic diseases. But even if cancer, cardiovascular and stroke deaths were completely eliminated, it has been estimated that life expectancy would only increase by about 15 years[8,9], and those 15 extra years would largely be spent in disability. People would then die from things that are currently put down to old age, including dementia. However, it seems likely that as obscure diseases became more important as the causes of death, more resources would be devoted to illuminating and overcoming them. This is why life expectancy has steadily increased over the last century even though the causes of death have changed. The resources of society are devoted to overcoming whatever are currently the main causes of death; when those causes change, the resources are reallocated to overcoming the next challenge.

Future death and disease throughout the world were predicted by the Global Burden of Disease Study (GBD) by the World Health Organization and Harvard School of Public Health in 1997, by projecting all known trends in different disease groups and populations[14]. They predicted that female life expectancy would reach 90 years by 2020 in developed countries. Increases in male life expectancy would be lower, particularly in developing countries, partly because of a dramatic increase in deaths due to smoking. The predictable consequence was more chronic disease and disability.

If maximum life expectancy is 100 or 120 years, is it really feasible for average life expectancy to approach and even exceed this figure in the future? Well, as we have seen, the concept of a fixed maximum life expectancy is dubious. If there is a maximum lifespan, it is probably increasing, although we can not be sure. There is no biological reason to believe in a fixed maximum lifespan – our genes have not programmed the body to self-destruct at a certain age – and even if they had, we could in principle override that programming. The body is after all just a complicated machine. There is no component of the body that could not be fixed or replaced, in principle, if not yet in practice. And there is no sociological reason to believe that our accelerating knowledge of medicine, biology and technology will not continue to deliver increasing lifespans, at least at the current rate of increase.

It may seem optimistic to believe that life expectancy will continue to increase at the same rate, but this is the current and past trend, so we should expect this to continue unless something else that we know about gets in the way. There is no reason to believe that 'maximum' lifespan will get in the way, any more than it did when the maximum was thought to be 65, 75, 85 or 100 years.

On the other hand, substantial further progress in increasing life expectancy and in particular healthy life expectancy beyond 100 years is going to require intervention in chronic diseases, dementia and the aging process itself. So it has been argued that a full-blown Tithonus scenario is unlikely because the chronic diseases and dementia that cause dysfunction in the aged will cause death once other causes of death have been eliminated. However, this is not inevitable, nor necessarily desirable. What is desirable now is that we concentrate resources on preventing chronic disease, dementia and aging, rather than preventing death – only then can we be sure of avoiding the Tithonus scenario.

Is there a cure for aging?

Is it possible to 'cure' aging itself, just like other diseases have been cured or at least tamed[15]? Cures for aging have been promised throughout history. The alchemists of ancient China, Hellenistic Egypt and mediæval Europe sought the Elixir of Life in various chemical potions. Hindus thought that controlling the breath, synonymous with the spirit of life, was the key to a longer life. The Romans adapted this belief in their own inimical way, suggesting that men feeling their age should sleep between two virgins, supposedly breathing in their youthful spirit. The same solution was suggested by Roger Bacon, the thirteenth century philosopher and scientist. Like Galen and Avicenna before him, Bacon believed that old age was due to the loss of innate moisture, and this could be restored by ingesting small quantities of substances, such as gold or 'bone from a stag's heart' that contained innate moisture or vital spirit.

Some have argued that aging is inevitable and immutable. However, a few years ago, a group of gerontology researchers, led by Aubrey de Grey, issued a manifesto asserting that, contrary to the current pessimism, it was feasible that, given sufficient will and money, aging could be cured and even reversed within a decade[16]. Wow!

They identified seven major forms of damage that accumulate in the body with age, and indicated current or foreseeable biotechnologies that could reverse each of these forms of damage. That is not just preventing aging, but reversing it. Indeed, they suggested that reversing aging may be easier that preventing it!

So according to these aging experts it is conceivable that a cure for aging could be found in the next 10 years, i.e. possibly in time to keep us alive 'forever'. If this were true: everything would change, both for us and the rest of human society. So clearly it is worth finding out whether the optimism of the gerontologists is justified. According to these gurus the seven steps to immortality are as follows:

Damage arising with age	Effects reversible by
Nuclear mutations	Delete the telomerase gene
Mitochondrial mutations	Insert mitochondrial DNA genes in nucleus
Cellular senescence	Destruction by immune cells
Aggregates in lysosomes	Insert bacterial genes that breakdown aggregates
Aggregates outside cells	Phagocytosis by immune cells; drugs
Immune and hormone decline	Hormone therapy; gene therapy
Cell loss	Stem cells; growth factors; exercise

The gurus don't think that nuclear mutations contribute to aging itself because the proportion of mutated cells is small. They do think that nuclear mutations cause cancer, and thus contribute to age-related disease and death. The solution they suggest is deleting the telomerase gene in all cells in the body, except germ cells and stem cells, thus preventing any budding cancer cell from turning on the telomerase gene to overcome the Hayflick limit and become immortal. This suggestion is certainly bold, but is not technically feasible at the moment, even in animals. A backup measure is to limit any cancers that appear using current cutting-edge medicine of either cancer vaccines or drugs that destroy the cancer's blood supply.

Step two on the road to immortality is to prevent or circumvent mitochondrial DNA mutations by putting a copy of the mitochondrial genome into the nucleus, where it is potentially less exposed to free radicals, better protected and more easily repaired. Again, this is no easy job: it would require a complete redesign of the genes involved, and then gene therapy that would deliver them to all cells in the body. This is a very tall order: unlikely to be possible in the next 50 years, if ever.

Step 3 is to destroy any senescent cells, that is cells that have reached their Hayflick limit and become moribund. The thinking here is that very few senescent cells arise during aging, but that this small number may be toxic to their neighbours, and therefore are best destroyed. However, it is not clear how the immune system could be induced to detect and destroy such cells. Nor is it apparent whether removing such cells would help.

The next steps involve molecular aggregates of one kind or another. Lysosomes are compartments within cells that digest defunct molecules or nasties such as bacteria that are engulfed by the cell. Lysosomes contain lots of enzymes that will breakdown the toughest materials. However, with age we accumulate within our lysosomes completely indigestible materials, particularly aggregates of free-radical-damaged fat and protein. The extent to which this indigestible material contributes to aging is unclear, but it does contribute to degeneration of the eye, and possibly arteriosclerosis. The suggested solution is gene therapy with bacterial enzymes that can break down more or less anything (including rock and toxic waste). The difficulty here, as elsewhere, is delivering the genes to every cell in the body and getting the genes to behave in a sensible way. Further, it would be impossible to find genes that would tackle every type of aggregate.

Insoluble aggregates also accumulate outside cells; for example, the β-amyloid plaques that characterize the brain of Alzheimer's patients, but also accumulate with normal aging. The suggested solution is to induce (by vaccination) the immune system to phagocytose the aggregates. This has been tried with Alzheimer's, but the results were equivocal, as the over-active immune system appeared to make things worse for the brain. Proteins outside the cell can also be cross-linked into aggregates by sugars such as glucose, causing an age-related decrease in tissue elasticity, which can impair the skin, muscle and blood vessels. The suggested solution is either stimulated phagocytosis of these aggregates, or current drugs that appear to break these cross-links and thus restore elasticity.

Cell loss with age in muscle, bone and brain may in principle be reversed by stem cells. That might involve isolation of stem cells from embryos or self, multiplication and differentiation in the laboratory outside the body, and then injection into the appropriate part of the body. Muscle and bone loss can be treated with appropriate exercise or growth hormones. Decline in function of the immune system with age may be treatable with appropriate hormones: a cytokine IL-7 in this case. Other aspects of aging may also be related to age-related decline in hormone secretion, which may be treated either by hormone administration or in the longer term by gene therapy to restore hormone release.

These seven-fold Strategies for Engineered Negligible Senescence (SENS) are being vociferously advocated by Cambridge polymath Aubrey de Grey. He and other gerontologists are seeking funding for a 10-year programme to block aging in mice. 'It will be total mayhem', says de Grey of the moment that people realize that mouse experiments suggest that immortality will soon be available:

It will cease to be just me saying this sort of thing: all my colleagues will be agreeing that serious extension of the human healthy lifespan is foreseeable. No one will want to stay in risky jobs. But the biggest thing of all is that people will behave as in wartime. It will be a true war on aging with everyone's first priority being to end the slaughter. It will mean voting huge expenditure to expedite the remaining science to translate the technology from mice to humans, but also voting even bigger expenditure to train the staggering number of medical personnel to administer the therapy when it arrives.

Aubrey de Grey has become a bit of a celebrity in both science and the media, relentlessly pushing his SENS agenda to funders, journalists and fellow scientists. Indeed, he has got up a lot of scientists' noses with his unorthodox approach. He sounds like a cross between an evangelical priest, second-hand-car salesman and charlatan peddling immortality. And he looks like Rasputin, with an enormous flowing beard and ponytail. But he is also brilliant at theoretical science, practical politics and getting things done. He has become the poster boy for the new immortality movement that he has done so much to generate. And this is somewhat incredible because until relatively recently he was just a computer technician in the Genetics Department in Cambridge.

He became interested in aging through talking to his wife, who is a research scientist. De Grey came up with a whole host of ideas about aging, and pushed them at aging conferences. Biologists (unlike physicists or chemists) are not used to the idea of theoretical scientists, and are generally unconvinced of their utility. But de Grey's engaging personality won some people over.

A few years ago I was an examiner for his PhD. A PhD is the professional qualification required to be a scientist, and it generally requires toiling away in a laboratory for three or four years. But the University of Cambridge has an unusual and rarely used route to such a qualification. Somebody who is already an established scientist can submit their previously published work or ideas for examination by two examiners, who proceed to dissect the work and its author in an oral examination. If the work is judged to be a substantial and original contribution to science the examinee gets awarded a PhD, and can wear a square hat and gown etc. De Grey was certainly an exceptionally talented examinee, and very entertaining too. But his unorthodox origins and approach have caused others to raise eyebrows. De Grey wants to stop aging by any means possible, as soon as possible, and is willing to do almost anything to achieve this aim. Thus he is more interested in medical engineering than basic science, and more interested in practical politics and fundraising than academia.

De Grey believes that if his SENS strategy adds 30 or 40 more years to current life expectancy, this will be enough for some people to reach immortality. That's because of his concept of 'escape velocity'. At the current rate of increase of scientific knowledge, the knowledge available in 40 years time will be sufficient to keep survivors alive for another 40 years, and so on, and on, and on.

Is this seven-fold path to immortality feasible? Recently, 28 leading researchers of aging responded to de Grey and denounced his SENS agenda as modern day charlatanism[17].

In the words of the great American journalist H. L. Mencken, 'for every complex problem, there is a simple solution, and it is wrong.' De Grey's research programme, which he terms 'strategies for engineered negligible senescence' (SENS), involves a combination of preventative and therapeutic interventions.... Yet, in his writings, de Grey fails to mention that none of these approaches has ever been shown to extend the lifespan of any organism, let alone humans...

Each one of the specific proposals that comprise the SENS agenda is, at our present stage of ignorance, exceptionally optimistic. Therefore, by multiplying the probabilities of success, the claim that all of these proposals can be accomplished, although presented with confidence in de Grey's writings, seems nonsensical. Consequently, the idea that a research programme organized around the SENS agenda will not only retard aging, but also reverse it – creating young people from old ones – and do so within our lifetime, is so far from plausible that it commands no respect at all within the informed scientific community...

Presented by an articulate, witty and colourful proponent, a flashy research agenda might catch the eye of a journalist or meeting organizer who is hunting for attention, publicity and an audience; however, the SENS agenda is easily recognized as a pretence by those with scientific experience.

Some parts of de Grey's agenda are feasible to test now; other parts are not currently feasible, but conceivable given enough resources; but for most of it (the first four steps in particularly) we have no idea whether it is feasible because we don't currently know how to do it. However, the worst problem is that, even if all seven steps were achieved, we don't know whether it would have any impact on aging at all, because we don't really know what causes aging in mammals.

Of course, if these interventions did slow aging, that would be good evidence that the corresponding mechanisms were responsible for aging. However, to get these interventions off the ground would require very extensive government funding, of the order of that for the Human Genome Project, but without a guaranteed end result. So we're in a bit of a chicken-and-egg situation: we don't want to invest in these potential solutions unless we know they have some chance of working, but we won't know whether they work until they are tried. That in a sense is why some gerontologists are trying to make such a big noise: they are attempting to bypass governments and normal funding mechanisms in order to raise expectations in the public so that they demand large-scale funding of research by governments.

There is no encouraging evidence that the promises and potions of these optimists are any more likely to work than those offered throughout history. No intervention of any sort has verifiably delayed the intrinsic rate of aging in humans. That includes exercise, diet, vitamins, antioxidants and hormones. What evidence we have suggests that these treatments have *no* significant effect on aging[18]. Calorie restriction delays aging in mice, flies and worms, but we don't know whether it would work in long-lived mammals such as humans, and even if it did it would be impractical to apply. To repeat, we don't know of *anything* that delays aging in humans, and we don't yet know how aging occurs. All else is promises and wild theories.

Unfortunately, there is no other choice than to test the wild theories of gerontologists – there is no other way to make progress. And we need to make progress fast, for the sake of both the individual and society, if we are to avoid the coming plague of disability and dementia.

The unravelling of death by the technologies of medicine

Technology is problematizing death. Technology has frozen conditions between life and death that had previously been considered in mythology, fantasy or philosophy. Until the advent of the respirator, the cessation of spontaneous breathing immediately led to the cessation of circulation and unrecoverable brain damage. Since the 1960s we have continually expanded the gray areas between life and death, stabilizing one process after another in the previously inexorable path from life to dust[19].

A variety of new and emerging medical technologies have started to unravel the historical concepts of death and self. New abilities to keep parts of the body or brain alive, while other parts are dead, have eroded the all-or-nothing quality of death. So rather than asking whether a person is dead, it becomes increasingly more relevant to ask: 'How much of the person is dead?' or 'Which parts of a person are dead?' The irreversibility of death, which appears essential to the concept, has also come under attack from resuscitation methods, brain repair technologies and the cryogenics movement. The very definition of death has become a battleground between various medical factions, and more worryingly the unified concept of self has become untenable.

In former times, collapse of any one of the vital organs (brain, heart, lungs, guts, kidney or liver) was followed within a few minutes, hours or days by irreversible death, depending on which organ was involved. Nowadays, that is not necessarily so, as various technologies have been devised to replace vital functions or reverse their collapse. The respirator can replace the function of failed lungs that can no longer breathe. The kidney dialysis machine can replace failed kidneys. The mouth, stomach and guts can be bypassed by drip feeding straight into a vein. Bones and joints can be replaced by mechanical equivalents. Mechanical hands connected to the nervous system are now being developed.

A completely failed liver cannot be replaced as yet, although some of its functions can, and it is likely that in the near future most functions could be replaced. In the short term (during heart surgery) the heart can be replaced by an external pump, but in the long term it cannot. However, this is likely to change in the near future, as devising a mechanical pump to replace the heart presents no overwhelming problems, and there are a number of current projects to do so. Specific functions of the heart such as pacemaking have already been successfully replaced. And a stopped heart can be successfully restarted by resuscitation or defibrillation, if you are quick enough.

Despite many technological advances the complete function of a failed brain still cannot be replaced (unsurprisingly). However, a dead brain no longer automatically means a dead body. The brain stem at the base of the brain controls the rhythmic contractions of the lungs and heart, so that death of the brain stem would normally stop breathing and circulation. But the respirator and control of circulation enables people with dead brains to be kept alive. How 'alive' is a hotly disputed question.

As an alternative to replacing organ functions by machines, the heart, kidneys, lungs, liver, skin, blood and blood vessels can be replaced by transplant from a human donor. Improvements in surgery and means of suppressing the immune system are now making possible the transplant of limbs: one American has had a functional hand transplant for the last five years. Even whole-face transplants are now feasible. This raises the thorny issue of whether you can really swap the body without affecting your identity or your 'self'. Would you really be the same self if you were transplanted into a new body or face? Surviving such transplants intact relies on the old concept that the self/soul is an indivisible and immutable substance, and resides entirely in the brain. However, this idea is becoming increasingly untenable.

We should also consider transplant transactions from the point of view of the donor. Most organ donors are 'dead' at the time of transplant, but the transfer means that their organs may live on in another host. Chronic shortage of human donors has led to the development of transgenic animals to act as more pliable donors. Pigs are particularly favoured as donors, essentially because their organs are of a similar size to humans. Genetic engineering of the pig is being used to reduce the immune rejection that follows transplant of any tissue into a foreign host. Further genetic engineering is aimed at making the organs more and more 'human' by inserting human genes into the animal's genome. This raises the spectre of farming chimeric animals that are part human, which has the advantage of making them more acceptable to the host: who wants the heart of a 'pig'? However, part-human animals will increasingly blur the ethical distinction between animals and humans, which may make the practice itself morally dubious.

We could avoid such moral dilemmas by growing organs in a dish starting from single cells. Skin can be grown in this way, and dish-grown skin is currently being used for transplant. Attempts are being made to grow other organs, such as liver and blood vessels, in a cell culture dish for potential transplant. However, it may be easier to regrow tissues in the body, rather than in a dish, if you introduce the right kind of cells into the body. This is the promise of stem cell technology.

Stem cells are cells capable of rapidly dividing and then changing (differentiating) into almost any type of cell. There are hundreds of different types of cell in the body, but once they have turned into their adult form they are incapable of turning into any other type. Of course all these cells were originally derived from a single cell type: the fertilized egg cell (the 'mother-of-all-stem-cells') that resulted from the fusion of sperm and unfertilized egg. And the embryo that develops from this fertilized egg cell contains stem cells. Actually the number of cells is very small, and at an early stage in development they differentiate into other cells, and so are lost. However, in 1998 James Thomson and John Gearhart in the US discovered that they could take differentiated germ cells from embryos, and treat these cells in such a way that they reverted to being stem cells of unlimited potential.

These embryonic stem cells can be isolated, for example from spare fertilized eggs at IVF clinics, and then grown in a laboratory dish. The cells can be differentiated into almost any cell type and then injected into the damaged organ of a patient, in the hope that they will then replace dead or damaged host cells and integrate with healthy host cells. For example, the adult heart contains mostly heart muscle cells, but it also contains neurons and other types of cell. If someone survives a heart attack, usually part of the heart completely dies. Some of the cell types that survive then divide and grow to replace those that were lost. Unfortunately, mature heart muscle cells are incapable of division, and the surviving cells cannot replace their dead neighbours. Medical scientists are now trying to use stem cells as a source of immature heart muscle cells that can be injected into the damaged heart, where they may divide and differentiate into mature cells to replace the lost muscle.

Medical technologies have contributed to the emergence of a new type of person in hospital: people who are brain dead but have a living body. People who had lost most of their brain function (often due to medical accident) but retained a functioning brain stem, and therefore normal breathing and circulation, were said to be in a 'persistent vegetative state'. Were these people dead or alive? Regarding such people as dead had certain strategic advantages, including conserving scarce medical resources, or the potential to use such 'dead' people as mobile organ donors and temporary incubators. But their friends and relatives might not be so easily convinced that such people really are dead, or at least irreversibly dead.

Three attitudes to the redefinition of death have developed in the United States, corresponding to three different criteria for death: neo-cortical death, whole brain death and whole body death. The neo-corticalists, starting with Robert Veatch in 1975, argued that people were dead once they had lost the ability to meaningfully interact with others, and the legal boundary of death should be set at the point at which a person entered a state of permanent unconsciousness. Consciousness was to be the criterion for life, and the seat of consciousness was the neo-cortex (the outer layer of the brain), so that death of the neo-cortex corresponded to end of the person. A variant of this position was advanced by Baruch Brody in 1988:

Wouldn't it be more appropriate to say that, even though (the permanently vegetative body) is still alive, this patient is no longer a person, having lost, when her cortex stopped functioning, the physiological basis of what is crucial to personhood?

In reaction to this radical new interpretation of death and personhood, many religious, political and medical conservatives maintained that the criteria of death should include both body *and* brain death. Eventually a compromise position was adopted whereby death was defined as death of the body *or* of the *whole* brain, including the brain stem. These criteria were written into the Uniform Declaration of Death Act (1980), which was enacted in 36 US states, declaring: 'Any individual who has sustained either irreversible cessation of circulatory and respiratory functions, or irreversible cessation of functions of the entire brain, including the brain stem, is dead'.

However, since then a number of developments have eroded confidence in the whole brain criterion for death. Dr Alan Shewmon, a neurologist at UCLA, demonstrated that some patients have 'survived' more than ten years of complete brain death. This showed that whole brain death did not inevitably lead to death of the body. In 1992, medics at the University of Pittsburgh approved a policy allowing terminally ill brain-damaged patients to be enrolled as 'non-heart-beating donors'. These included patients who were not 'dead' by the whole-brain criteria, but had brain injuries sufficient to require a respirator. They were rolled into the operating theatre, the respirator was turned off for two minutes, and then organs were removed.

However, in 2003 Terry Wallis awoke from 19 years in a comatose state close to the persistent vegetative state, dramatically demonstrating that such states were not irreversible. This and other cases have led to the introduction of a new diagnosis of 'minimally conscious state', where some level of consciousness is retained and may be reversed to a close to normal level of consciousness. This is in contrast to the persistent vegetative state, where there is no level of consciousness and no reversal. However, the distinction between the two diagnoses has been criticised as being arbitrary, as in practice a continuum of levels of consciousness is found in patients, ranging from a complete absence to complete presence of consciousness. Furthermore, different patients with partial consciousness have different aspects of consciousness present or missing, such as particular senses or mental abilities.

The brain death criterion of death has appeared scientifically plausible partly because the death of the brain's most important cells, neurons, and thus brain damage generally, has been thought to be irreversible. Mature neurons are said to be 'post-mitotic'; that is, they can no longer divide to produce new neurons, which could replace dead ones. In contrast, most other cells in the body remain capable of dividing to replace dead cells, even in aged individuals. During fetal development, cells migrate all over the forming body and differentiate into cell types appropriate for their position in the body. This differentiation into a particular cell type is normally irreversible, and it was thought that in the mature body there were no cells left that could differentiate into neurons. Hence when a neuron was lost, it was not replaceable.

However, research over the last five years or so has overturned a century of belief to show that not only can embryonic stem cells differentiate into neurons, but also the mature brain retains a small number of cells that are capable of dividing and differentiating into neurons: neural stem cells. Surprisingly, it is possible to inject neural stem cells into the brain and watch them migrate to damaged areas, where they differentiate into neurons and other types of brain cell. This would have seemed like ridiculous science fiction only a few years ago, but now it is practical reality.

However, it is still far from clear that such stem cell therapy will aid damaged brains. The brain is not just a conglomeration of cells – some scientists consider that it is the most complex structure in the known universe. The brain contains roughly 100 billion neurons, of many different types, each neuron being connected to roughly a thousand other particular neurons. And the particular arrangements of the connections matter, as these direct the flow of electrical signals through the vast network of neurons. The particular connections and patterns of electrical signals constitute the information processing of the brain, just as they do in a computer. Hoping to repair a damaged brain area by injecting it with cells is like hoping to repair a damaged computer by throwing transistors or chips at it. It is not just the presence of transistors or neurons that matters, but the mind-bogglingly complex arrangement of those units on centimetre to nanometre scales is also critical to function. However, most of these connections have to be assembled during development according to some genetic plan and then sculpted and tweaked by experience. Thus throwing neurons at damaged brains may do more harm than good.

On the other hand, there are some types of brain damage where the organization of neurons may not be so crucial. For example, patients with Parkinson's disease lose large numbers of a specific type of neuron (neurons that release dopamine) from a particular part of the brain. And in principle these neurons could be replaced by neural stem cells differentiated to produce dopamine. That's because the spatial and temporal pattern of dopamine release is not as important for the paralysed Parkinson's patients so much as the mere fact that some is produced. Embryonic stem cells are extracted from aborted human foetuses, multiplied and partly differentiated in a dish, and then injected into the damaged area of the brain. The patient is unlikely to care that part of her self is being changed by this procedure, because the disease itself has radically altered her self. But should we worry if this type of procedure is used, as proposed, to protect memory, enhance intelligence or reverse brain aging?

The last decade has also seen huge strides in our ability to genetically manipulate cells to get them do almost anything we want. Stem cells are now being genetically engineered before being injected into the body or brain, in order to help them turn into particular types of cell or to direct them to particular areas. Gene therapy is also being used to cause expression of genes in the brain that favour repair. For example, genes have been introduced into the brain that direct the synthesis of Nerve Growth Factor, which is essential for the growth and survival of certain types of neuron. The study and practice of brain repair and restorative medicine are new academic and medical subjects. And there are many new institutes, departments and companies being devoted to them. However, it should be noted that very few practical, helpful medical treatments have been developed so far.

Brain repair is also being approached using electrical engineering. If electronic devices (memory chips, computers, digital cameras) could be married with neurons the potential would be enormous. But first we have to find a way to connect electronics to neurons using electrical or chemical signals. The National Institute of Health (NIH) is funding a program of research on neural prosthesis to develop such neural-electronic interfaces for medicine. Silicon chips have been developed with an open lattice structure coated with a pattern of biochemicals that encourage the growth of discrete connections with nearby neurons. And these chips use neural network software, originally inspired by neuronal communication, to communicate with the neurons. For example Dr Roy Bakay at Emory has implanted such a chip plus radio transmitter in the brain's motor region of a completely paralysed patient, and taught him to control a computer's cursor by thinking about motor actions. This generates electrical signals in the motor cortex that are picked up by the chip and broadcast by the transmitter to a nearby computer. The computer then interprets the signals, and in this case directs the movement of a cursor on the computer screen, but it could do other things.

Matt Nagle has a chip implanted in his brain enabling him to control a robotic arm. Matt is paralysed from his neck down due to a vicious knife attack, so he volunteered for an implant known as BrainGate, developed by John Donoghue, professor of Neuroscience at Brown University on Rhode Island, USA. A hole was drilled through Matt's skull, and the BrainGate chip pressed onto the surface of his brain, so that 96 hair-thin electrodes, each a millimetre long, penetrated into his motor cortex. This area of the brain would normally control muscles in the body, but since Matt's spine is severed they don't. However, the neurons in his brain still fire when he wills his arm to move. So the electrodes in the chip pick up this electrical signal and transmit it to the robotic arm and hand. Matt's ability to control the arm, simply by thinking about it, is still limited – limited by the tiny, tiny fraction of the brain's activity that the electrodes are sampling – and limited too by the inflammation that forcing 96 electrodes into the brain causes. Still, that little bit of control means a lot to Matt. And he has also been hooked up to a computer so that he can turn on and off a TV, and change the channel and volume – just by thinking about it.

Other ways of connecting to the brain, without sticking electrodes into it, are also being devised. Scalp electrodes have been built into a kind of cap, which when worn registers global activity from the brain. This gives a much cruder measure of brain activity, but surprisingly people wearing the cap can still control a computer cursor by thought alone. These types of interface are likely to improve, potentially making it possible to connect the brain to memory chips, sensory devices, computers or remote devices such as the World Wide Web. However, it is still unclear whether electronic devices could replace the information-processing roles of parts of the brain. And the challenge of interfacing electronics with the brain remains a major stumbling block.

Convergence between the technologies of electronics and biology is occurring at a number of different levels. This convergence, if it goes more than skin deep, has a huge transforming potential. The revolutionary changes that have occurred in the last generation in our power to manipulate electronics and biology have outstripped almost all other human endeavours. But only now is the interface between these two realms being seriously explored.

Nanotechnology is a term that covers many disparate fields attempting to engineer devices or processes at or approaching the nanometre scale – one billionth of a metre. In many of these fields biology is the subject or inspiration of this engineering, because billions of years of evolution have perfected a plethora of nanometre-scale biological devices, such as the thousands of protein machines that make our bodies and all cells work. Many of these machines work on electricity. For instance, our mitochondria have a five-nanometre molecular device that transforms the energy in our bodies from an electrical form into a chemical form (ATP). It consists of two motors: one couples a flow of electricity through the device to the rotation of an axle; the second couples the rotation of the axle to the synthesis of ATP.

A number of laboratories are genetically engineering this and other molecular machines to produce electrically or chemically driven gates or propulsion devices. Part of the aim of such research is to construct devices that could be put into the body to repair damage or prevent disease. Such devices could be inserted into the body either by gene therapy (getting the body to construct the devices by supplying the DNA blueprint), or by injection of re-engineered cells. Alternatively, these microscopic machines could be used to construct macroscopic devices that would replace the function of diseased or dead organs. Nanotechnology is trying to bridge the gap between biology and electronics – but this remains a big gap.

All these technologies and potential technologies challenge our concepts of death and self. If we can survive with large parts of our body and brains dead, then death cannot be an all-or-nothing event. And if large parts of our bodies and brains can be replaced by machines, devices or biological spares, and the mind can be augmented and hooked up to the Internet, then the self cannot be a unified all-or-nothing thing. We may have to get used to the ideas of parts of the self, multiple selves, selves fused with machines, or selves shared over a network.

Life beyond death?

The potential future reversibility of body and brain damage was the inspiration for the cryonics movement in the US: the idea of freezing recently dead people in the hope that their bodies and brains may be repairable in the future. The idea of preserving the body after death for future resurrection is very old – the Eygptians tried mummification, Benjamin Franklin suggested pickling the body, Mary Shelley imagined reanimating the body with electricity.

Currently about 120 'dead' 'patients' are cryogenically frozen, and about 1000 live patients have signed up to be frozen after death. With today's technologies it is hard to see how the damage inflicted by the freezing of cells could be reversed, as ice crystals effectively turn all cells in the body into mush. You might as well try to revive a pile of mud or a clay figure of a person as try to revive a body freeze-thawed into mush. However, these people are not banking on today's technologies to repair them, but rather those of tomorrow or the day after, which cannot be predicted. The 'cryonauts', as some like to be known, are voyaging into the future. If these bodies and brains could be repaired it still remains a matter of debate whether the revived person would be the same person as previously died. A *sine qua non* of such continuity of identity would be the presence of memories, at least from the time of death. We do not know whether memories could survive this process, as it is still unclear in what form such memories are stored.

A major boost for the cryonics movement was the publication of Eric Drexler's book *Engines of Creation* (1986). This provided a vision of nanotechnology, but also included a discussion of the feasibility of using nanotechnology to repair the frozen ice-damaged bodies of cryonauts. However, at the moment his vision remains largely science fiction rather than science fact.

Future technologies are difficult to predict. Only a few years ago the idea of cloning a human would have seemed science fiction. Now cloning can be done almost routinely with many mammals, and if we really wanted to, it could almost certainly be done with humans. Cloning requires only the nucleus of a single cell from the donor's body. The nucleus is the part of the cell containing the DNA. It is likely that within a few years such a nucleus could be extracted from a frozen cryonaut and injected into a human egg cell from which the normal nucleus had been removed. The resulting hybrid egg cell would be inserted into a surrogate mother's womb and there grown into a clone of the cryonaut. Would any self-respecting cryonaut accept such a procedure as a legitimate rebirth? Would she think her contract with the Alcor corporation to revive her had been fulfilled (and worth every penny)? Would the clone, when adult, think of herself as having a continuity of identity with the cryonaut?

At first glance it would appear naïve to think of a clone as having a self or self-identity continuous with the clone's nuclear donor, even though the clone and donor are genetically identical. Some people feel a continuity of identity with their parents or children, but this may seem a rather abstract or metaphorical continuity. Identical twins are genetically identical in the same way as clones, and can be regarded as clones of each other, but do not appear to have identical or continuous selves. Differences in environmental conditions and life experience produce remembered experience, learning and emotional conditioning that shape the self differently in different people. Thus both clones and twins do not have identical selves with their genetically identical counterparts.

However, the fact that they do not have *identical* selves does not rule out the possibility that they have similar selves or parts of their selves in common. Many studies have shown that identical twins, even those brought up in radically different environments, have remarkably similar psychology, personality, beliefs, behaviour, intelligence and feelings. Although they do not have identical selves, parts of the self may be more or less identical, whereas other parts differ. In addition, particularly during childhood, identical twins often form an exceptionally close relationship with each other, such that for example they can finish each other's sentences. In these circumstances it is tempting to think of twins as sharing a common identity, or even sharing a common self.

Is a self the kind of thing that can be shared? Is a self the kind of thing that can be partly identical in different people? Is a self the kind of thing that can be passed from one person to another? Of course, as we have discussed before, it all depends on what we mean by 'self'. Our culture has inherited a concept of self as an indivisible and immutable substance: the soul. The unchanging nature of self is difficult to reconcile with what we have already accepted, i.e. the self is partly shaped and changed by upbringing and life experience. I am not the same person as I was when I was born, when I was 1 year old, when I was 10, when I was 20, or even to a very limited extent as I was yesterday. I am not the same person today as I will be when I am very old, or when I am dead. There may be bits of me that are more or less identical with me in the past or the future, but there are also bits of me that are different. If my one-year-old self could be brought into the same room as me today, how much self-identification would I have with this person? The apparent continuity of self from day (via extinction at night) to day, is fostered by the apparent continuity of memory and the very slow rate of change. If a dramatic or traumatic experience causes an abrupt change (e.g. a religious conversion) then we may think of ourselves as a new person, discontinuous with our remembered selves. In these circumstances, religious scholars would distinguish between the self as soul, which is immutable, indivisible and immortal, from the self as spirit, mind or psyche, which is potentially changeable, divisible, and mortal.

The potential fallibility of memory has led some to suggest that immortality or very long life is an illusion, because someone who was 1000 years old could not possibly remember their whole life, so their life would consist of a series of different selves. In these circumstances, the advantage of immortality or very long life is unclear. Others have argued that even if memory were insufficient to span the whole of life, the continuity (as in normal length life) is sufficient to justify bothering with immortality.

If the clone of a person differs from the original person on the basis of its life experience, would it be possible for the clone to acquire those life experiences in order to reconstitute the original person? This is essentially the dream of the Raelian Movement. This UFO cult believes in the scientific realization of immortality in two (not-so-easy) steps. Step one is to clone the potential candidate for immortality: the cult claimed to have cracked this at the end of 2002, with the (virtual) birth of the first human clone named Eve. Step two is to download all the experience from the candidate's brain into the clone's brain.

Again, at first encounter it seems unlikely that one could suck up all the life experience of a person into a syringe and inject it into another person's brain with any chance of transferring the experience. At second encounter it doesn't seem so obvious either. However, there is a tried and trusted method for passing on experience from one person to another: it's called communication, and it is the basis of human culture. As well as passing on their genes to their children, parents pass on their beliefs, attitudes, behaviours, tastes, religion, stories and memories. By talking, teaching and training, by reading, watching and copying, culture is transmitted from one person to the next, and from one generation to the next. The units of culture that are thus transmitted are known as 'memes'.

A meme could be an idea, a method or a catch phrase: anything that can be transmitted from one person to another by communication or copying. A meme can reside in a book, a film or a person's brain. I am transmitting the concept of memes from my mind to yours via this book. The concept of memes was originally popularized by the Oxford evolutionary theorist Richard Dawkins, who coined the term 'meme' in analogy with 'genes' to point out the close parallels between cultural inheritance via memes and genetic inheritance via genes. Memetics, like genetics, has rapidly spread through scientific disciplines (although social scientists in general do not like this concept). The relevance of memes here is that memes, together with genes, could constitute the self (as spirit, mind or psyche). You are the sum of your beliefs, training and learned behaviours, together with your genetic programming and innate behaviours.

Modes of survival

Traditionally there are three ways of surviving death: spiritually, genetically or culturally. That is, as a soul, a descendant or a meme. These supposed modes of immortality have been immensely important motivations within society. Without them our cultures and society would be unrecognizable.

Civilization as we know it would not have existed without these three forms of immortality dangling in front of us.

Spiritual survival is survival after death, generally without the previous body, as some form of coherent self, variously termed spirit, soul, ghost or psyche. Most cultures and religions have or had some such belief. The extent of supposed mental continuity before and after death is not always clear, but is usually large, including memory, character, beliefs etc. Religions claim exclusive knowledge as to how to survive (or survive better) spiritually. Obviously the promise of spiritual immorality or spiritual improvement has been an important motivator in society historically, and remains so today.

Genetic survival is survival of the genes through one's children and one's children's children. Again this has been a huge motivating factor in society. It is the basis of the family unit, both in animals and humans. People do extraordinary things for their relatives, children and children's children. Mothers, both human and animal, will lay down their lives in defence of their children. Fathers have devoted their lives to promoting the family's interests. Heredity, blood, family name, clan honour and ancestor worship have been key motivating factors in all kinds of societies and times. Are these things done in order to gain a piece of immortality? Clearly they are not done to gain immortality for the self. No one believes that by having children one can survive death as a self. But equally, by having children and promoting their interests, one is ensuring survival of the family, name and genes. Much of the time this is done unconsciously, but often there is a sense of being something other than an individual self, of being part of a family with a history and a future. Sometimes there is an explicit feeling of contributing to something greater and longer lasting than an individual self.

Cultural survival is survival of our works, deeds and ideas within human culture. We might gain this form of 'immortality' by writing a great novel, thinking up a successful scientific theory, designing and building a house, killing a lot of people or becoming famous. Again this form of immortality has been a great motivator throughout history, perhaps the greatest motivator driving people to contribute to society. It works both on the grand scale of great artists, scientists and politicians doing immortal things, and just as importantly on the milder level of being nice to other people, making or doing things that last. People contribute to society for all kinds of conscious and unconscious reasons, but if their contribution lasts or is reproduced, then a little bit of them survives, independent of the self or genes, within the culture. Some would call this a 'meme'. It might be a small thing like something you said that someone remembers. Or it might be something big, like a grand theory of everything.

So immortality can take at least three different forms: spiritual, genetic and cultural, corresponding to the survival of the soul, genes or memes. And paradoxically these modes of immortality have motivated individual selves to contribute to society. Of course people do things for all kinds of reasons apparently unrelated to seeking 'selfish' immortality. But what counts as selfish depends on our definition of self.

Spiritual survival, survival of the soul, has no scientific basis. It is an ancient concept which, like the concept of angels dancing on pinheads, is not compatible with our present scientific understanding of the world. Despite the deep desire that we and our loved ones should survive, any honest survey of the facts will show that there is no verifiable evidence of survival of the spirit/soul/mind after death. Not one of the billions of people who have died in the past has come back to talk to us or otherwise interact in any meaningful way. How then do we account for the fact that so many apparently sensible people, including scientists, believe in ghosts, souls and Father Christmas?

Proving that the soul does not exist is not easy. As Karl Popper pointed out, you can't logically prove that something does not exist. There is an asymmetry of proof: a single observation of a blue swan would be enough to tell me that blue swans existed, but not seeing blue swans a million times would not be sufficient to show that blue swans did not exist, as they might exist somewhere I had not yet looked. So it is logically impossible to prove that something does not exist, unless the concept of that thing contains some kind of internal contradiction, such as a married bachelor. This asymmetry of proof allows us to hold onto all kinds of appealing beliefs for which there is no positive evidence, such as the existence of angels, gods, dragons, ghosts, elves, Father Christmas and souls.

In practical terms, though, we accept that something does not exist if we (and all credible witnesses) never observe it. However, the truth of a belief is not necessarily the only good reason for holding that belief. It may be sensible for you to believe in something even though it does not exist, if believing it makes you happy. Thus if the non-existence of immortality, god, human goodness, or the greatness of the English cricket team made you chronically sad, you would be stupid not to continue believing these things, as long as this was psychically possible. Many apparently sensible people have made themselves miserable by only believing in things that were true. How stupid can you get! Indeed, there is evidence that depressed people generally have a more accurate view of the world than happy people (see Kay Jamison's book *Exuberance*). The implication is that having an accurate view of the world could make you depressed, whereas seeing the world through rose-tinted glasses may help make you happy. But how can you believe in things that are untrue? As the White Queen informed Alice (in *Through the Looking Glass*), believing impossible things is not itself impossible, it just takes practice.

Genetic and cultural survival are combined in having children and then bringing them up. You can imprint both your genes and your memes on your children. One half of your genes are passed on to each of your children (you have two sets of genes, only one of which goes into the sperm or egg cells), but a randomly different half of your genome goes to each child. So having a few children will ensure the survival of most but not all of your genes. A child's upbringing is one of the most important influences on a person's memes, core beliefs, ethics, outlook, and behaviour patterns. The family is an institution designed to transfer genes and memes to the next generation, and ensure their survival.

However, while gene/meme transfer in the family is fairly deep and secure, it is also of relatively limited extent in terms of the number of people affected. To ensure wider dispersal of the genes and particularly the memes we must look beyond the family to the wider community. Women and men are thought to have different strategies for ensuring survival of their genes. Women can have relatively few children in a lifetime and must invest substantial resources in carrying each child, while men can have many children and invest very little in each. According to this view, women after birth continue to invest heavily in ensuring the survival and benefit of their children, whereas men are less concerned with individual children and more interested in simply spreading their genes as widely as possible. Women invest deep and narrow, while men spread themselves shallow and wide.

Or at least that's the Evolutionary Biologist's story. It certainly works well in describing the relative gender investment in childcare in most animals. But does it work so well in humans, who are meme machines as well as gene machines? The modern tendency towards a more equal gender investment in childcare suggests either that the genetic differences were not so large or that culture can overcome genetics. On the other hand, the more equal investment in childcare may simply reflect more monogamous relationships in the modern world or the widespread use of birth control, so that because men now have a similar number of children to women, they are now motivated to invest in their upbringing to a similar extent.

The traditional means of genetic survival is through having children, who inherit half your randomly selected genes and half of your partner's genes. In future it may be possible to dramatically improve on this by cloning yourself, and thus ensuring complete (or almost complete) genetic survival – almost complete because mitochondrial DNA is not transferred to the clone. However, mitochondrial DNA is a very tiny fraction of the total, and men never transfer it to their children anyway, as it is always inherited from the mother.

Why then, if cloning is such a dramatically better way of ensuring genetic survival, is it currently so unpopular? Partly this must reflect the conservatism of society, and the implied threat to the family as the basic unit of society. But the current revulsion for cloning is based on the idea that it is 'selfish'. This is a somewhat odd response of liberal, capitalist society, which is meant to respect selfish individualism as a fundamental right and the engine of economic growth. Further it is hard to see how wanting to clone oneself is any more selfish than wanting to have children or wanting not to die. Of course we can object to it on the basis that nasty people might do it, but this is like objecting to mobile phones because nasty people might have them. Whether cloning is indeed selfish depends on whether the self is in any sense transferred from the donor to the clone. This in turn depends on our concept of the self. If we think the self is unitary, immutable, indivisible and monolithic then it cannot be transferred to the clone. But if, as I believe, the self is a mutable community, there is a sense in which the components of the self can be transferred as genes and memes to our children and clones. Whether this is a good or bad thing I don't know; probably it has nothing to do with good and bad. On the other hand, a society full of clones is potentially a fractured society, as clones might potentially only seek to benefit their own co-clones.

Memetic proliferation is becoming ever easier in the modern digital world of global media and internet. But paradoxically the survival of memes is threatened by the same technological trends, as every idea is drowned in a sea of information and the insatiable desire for the new. So the new becomes the undesirable old five minutes later. Every news corporation, fashion house and publishing house is looking for the 'news' rather than the 'olds'. Even in science the 'news' is what matters. Of course there's an up side to the rapid spread of memes in modern culture, in that we get a hell of a lot more memes. And we get the memes that we want rather than the memes that our parents or local vicar wanted us to have. And if we do create memes there are many more ways of reproducing and spreading those memes around the world.

New media technologies have been applied to the ancient business of securing the memetic survival of dead people. *In memoriam* websites have sprung up all over the web, recording and celebrating people's lives. New age cemeteries in Hollywood include electronic archives of the deceased: with a CV, testimonies from friends and relatives, and videos of the deceased (before death). The undertakers can supply a film crew to record the autobiography (before death) or biography (after death). In a rather different direction, a 'natural death' movement has started in analogy with the natural childbirth movement. This movement encourages people to manage their own deaths at home in a non-medicalized environment; helps the family to dispose of the body and perform the 'funeral' without the interference of 'professionals'; and promotes environmentally friendly modes of body disposal, and non-traditional cemeteries and services.

How can you directly transfer your memes to a clone or other person, to ensure full memetic and genetic survival? This isn't easy, and isn't likely to be available any time soon, if ever. But the most promising approach is to develop extensive electronic-neuronal interfaces in the brain, by the methods discussed above. For example, neurons can be induced to grow on the contacts of microchips, and the electrical signals used by the neurons can be read by the microchips. The chips could potentially also signal back to the neurons using electrical or chemical signals. Neuron-interfacing chips can and have been implanted in the brain and connected to a radio-transmitter, so that electrical signals in the brain can be read. Moreover, humans have been trained to directly stimulate these chips in their brain, by their thoughts changing the electrical signals in the neurons, which in turn change the electrical signals on the chip.

The radiotransmitter attached to the chip can be used to control some external electronic device, for example a computer pointer, so a paralysed human can control the pointer by thought alone. If signals are fed in the opposite direction, then in principle external signals transmitted to the chip could be detected by the neurons, so feeding information, experiences or commands directly into the brain. As the Canadian neurosurgeon Wilder Penfield showed back in the 1940s, a vast range of subjective experiences can be evoked by electrically stimulating different parts of a patient's exposed brain. So in principle it is possible to evoke defined experiences in brain matter, but in practice sticking electrodes and chips into people's brains is not going to be very easy or popular.

Neural-electronic interfaces are still in early development, so that we still don't know what is going to be possible. But already people are proposing that memories or knowledge might be transferred to computer chips locally or remotely. Implantation of a memory chip might help people with failing memories. Or working in the other direction, we might learn Chinese by implantation of a Chinese language chip. Now if this were possible, it is also conceivable that someone might download their thoughts and memories to a computer chip or network that was then plugged into a clone of themselves. However, at present it seems that memories are not the kind of thing that could be transferred to a computer chip. That hasn't stopped people imagining, for example, people's brains connected to remote robots or the Internet, or (as in the *Matrix* films) a computer network directly controlling people's brains and experience.

Having considered genetic and memetic survival, we need to reconsider spiritual survival. What is spiritual survival? At least two answers have been given, depending on two different concepts of the soul. Firstly, there is survival of the self without the physical body, including individual personality, experience and continuity with the pre-death self. This type of survival appears impossible, because we know that experience and memories etc. are produced by complex neuronal processes in brain matter. They might conceivably be reproduced in computers or other matter, but it is not currently conceivable that complex entities can exist in nothing, with no matter.

However, there is another concept of soul that is not so easily dismissed. Both modern and ancient theologians have argued that the soul is that part of the self that is divine and immortal. This goes back to the ancient idea that soul, spirit or pneuma is breathed into the body at birth, and taken back with the last breath of life. According to this concept, soul is not a self but rather a constituent of ourselves – it's not an entity but rather a substance or principle that is immortal. Now we know of several constituents of ourselves that are

immortal, or at least less mortal than our bodies and selves –
these include our matter, our memes and our genes. We could
– at a stretch – regard these as our true souls.

However, we would be missing something essentially personal and subjective from the concept of soul.

A better candidate for the modern soul would be consciousness itself. There is an important sense in which all that exists is
consciousness. And the only thing that ties consciousness to
the individual is memory. Without memory we could regard
each instance of consciousness as separate – each flicker of
subjective experience a world to itself. Alternatively, it is open
to us to regard all consciousness as one (the divine?). However, memory ties together particular experiences. Within a
dream I identify with the dreamer – I am the dreamer – no
matter how different from my waking self. If I remember the
dream when I wake, then this piece of world consciousness is
annexed to me. If I forget the dream then it floats free as an
independent bubble of consciousness. In this sense we can
regard our entire lives as a dream – a bubble of divine consciousness. However, we will not wake from this dream at
death – just the opposite: at death we will fall into a dreamless
sleep. But there will be other dreams in other people and
places. Whether we identify with those other dreams is a question of... identity. We can choose to identify or not to identify,
just as we can choose to identify or not to identify with ourselves as babies, our future selves, our children, our community or our god. Of course not all identity is open to free choice
– it's partly determined by our genes and our upbringing. But
that part of identity that is open to choice can be shaped by
our concepts – such as our concepts of self or death.

Assuming that we could avoid mortality, would we like our
immortality to be permanently the same or permanently
changing? It's an old dilemma. The ancient Greek philosopher
Heraclitus maintained that *everything* changes, while his suc-

cessor Parmenides retorted that *nothing* changes. Many would feel that an immortality where we and everything else stayed the same – where we did not grow, learn, change – would be Hell. But if we were permanently changing, could we really count this as immortality? To some this is a justification for death – to keep things changing and prevent them stagnating. To others it is the reason to celebrate life as change – and to keep on living as long as possible, while changing.

postscript
and how should we die?

Dante and Virgil, having faced death and plumbed the nine depths of Hell, finally emerge to see the stars once more. And we too have survived nine chapters of death and desolation, hopefully to emerge with our concepts of death and self reformed. But having used up our nine lives, perhaps it's time briefly to consider what to do about death. I would suggest nine things to do before we die:

1. **Rage, rage against the dying of the light**. The deaths we face today are unnatural; we should not accept them as our inevitable fate.

2. **Make the last 10 years of life worth living**. As a society we should stop turning a blind eye to the multiple miseries at the end of life. Let's attack the diseases, disabilities and dementia full on. And let's reintegrate the aged into society or whatever society they want.

3. **Accept that death is an analogue process rather than a digital event**. Death is part of life. If we realize this, we can attack death within life, and value life after death, rather than being preoccupied by the terminal event.

4. **Think of our selves as waves rather than atoms**. If we accept this, we may treat the self as a delicate and diverse entity that requires protection and nurturing. We are part of society, and society is part of us. We should not be so concerned about who we were, or who we will be, but rather who we are now and what we are doing now.

5. **Fund research into aging and dementia**. Current research funding is almost derisory. Many diseases have been cured or ameliorated by medical research. If we care about aging and dementia let's do something about it.

6. **Make medicine about maximizing life, not preventing death**. Current medicine and medical research is hung up

on preventing death, turning acute death into chronic death. We need to change research funding and medical practice to prioritize extending healthy life rather than life at any cost.

7. **Stop turning acute diseases into chronic diseases.** Revamp the economics of the pharmaceutical and biotech industry to stop them benefiting from the extension, rather than the cure, of chronic disease. Overhaul the clinical trial and patenting system that prevents medical advances.

8. **Face the fact that we are going to die and prepare for it.** Let's stop pretending or ignoring it. We need to find out how people want to die and enable them to do it. Euthanasia will be inevitable for some people – let's do it properly. Stop sending people to hospital to die. Hospices should be as ubiquitous and well funded as maternity hospitals.

9. **Live life to the full – spread your memes and genes.** Stop living life as if we were immortal – start living life as if it were going to end soon – it is. Seek out immortality, wherever it can be found, but not at any cost. Immortality as a unified self is not currently achievable, but that does not mean we should stop trying, and there are other ways of reaching immortality that are achievable now. Leave something really worthwhile behind you – build that dream, write that novel and... have lots of sex.

references

Chapter 1: Beginnings and endings

1. From 'Do not go into that good night' by Dylan Thomas, in *The Collected Poems*, New Directions Publishing Corporation, New York, 1952.

Interlude 1: A brief history of death and damnation

1. *The Divine Comedy: Inferno*, Dante, c1320 Translated by Mark Musa, Penguin Books, London, 1971.

2. Translated from Homer by E. V. Rien, *The Odyssey*, Penguin Books, London, 1946.

3. Virgil, *Aeneid*, translated by David West, Penguin Books, London.

4. Milton, J. (1667) *Paradise Lost*. Penguin Classics, London.

Chapter 2: The changing face of death

1. Austad, S. N. (1997) *Why We Age*. Wiley, New York.

2. Source: US Census Bureau.

3. Riley, J. C. (2001) *Rising Life Expectancy: a Global History*. Cambridge University Press, Cambridge.

4. Source: *A Comparative View of the Mortality of the Human Species* ... (London, 1788), and John Charlton & Mike Murphy (eds.) *The Health of Adult Britain 1841–1994* (London, 1997). Office of National Statistics (2004) *Death registrations in England and Wales, 2003: Causes report of Health Statistics Quarterly*. The Stationery Office, London.

5. Prior, L. (1997) Actuarial visions of death: In *The Changing Face of Death* (eds. Peter Jupp and Glennys Howarth). Palgrave Macmillan, Basingstoke.

6. Bauman, Z. (1992) *Mortality, Immortality and Other Life Strategies*. Polity Press, Cambridge.

7. Gray, A. (2001) *World Health and Disease*, 3rd edn. Open University.

Chapter 3: How we die today

1. http://www.heartstats.org/

2. Stevenson, L. W. (2005) *European Journal of Heart Failure*, **7**, 323–31.

3. National Heart, Lung and Blood Institute (1998). *Morbidity & mortality: 1998 chartbook on cardiovascular, lung, and blood diseases*. US Department of Health and Human Services, Rockville, Maryland; National Institutes of Health.

4. Weir, H. K., Thun. M. J., Hankey, B. F., Ries, L. A., Howe, H. L., Wingo, P. A. *et al.* (2003) Annual report to the nation on the status of cancer 1975–2000, featuring the uses of surveillance data for cancer prevention and control. *Journal of the National Cancer Institute*, **95**, 1276–99.

5. Parkin, D. M., Bray, F. I. and Devesa, S. S. (2001) Cancer burden in the year 2000. The global picture. *European Journal of Cancer*, **37**(suppl 8), 4–66.

6. Cancer Research UK: http://info.cancerresearchuk.org/cancerstats/

Interlude 3: The search for immortality

1. *The Epic of Gilgamesh*, translated from the Sumerian by Andrew George, 2000. Penguin Classics, London.

2. *The Epic of Gilgamesh*, translated by N. K. Sandars, Penguin Classics, London.

Chapter 4: Death is falling apart

1. Office for National Statistics, UK, 1998.

2. Federal Interagency Forum on Aging-Related Statistics. *Older Americans 2000: Key Indicators of Well-Being.* Federal Interagency Forum on Aging-Related Statistics, Washington, DC, U.S. Government Printing Office. August 2000. Sources: Current Population Reports, *Americans with Disabilities, 1997,* pp. 70–3, February 2001 and related Internet data; Internet releases of the Census Bureau and the National Center on Health Statistics).

3. http://www.chcr.brown.edu/dying/

4. Office of National Statistics (2000) *The Mental Health of Older People.* Stationary Office, London.

5. Seale, C. (2000) Demographic change and the experience of dying. In: *Death, Dying and Bereavement.* Sage, London.

6. Federal Interagency Forum on Aging-Related Statistics. *Older Americans 2000: Key Indicators of Well-Being.* Federal Interagency Forum on Aging-Related Statistics, Washington, DC, U.S. Government Printing Office. August 2000.

7. Addington-Hall, J., Altmann, D. and McCarthy, M. (1998) Age and Ageing, **27**, 129–36.

8. Lawton, J. (2000) *The Dying Process.* Routledge, London.

Interlude 4: The immortals of Luggnagg

1. Swift, J. (1726) *Gulliver's Travels.* Oxford World Classics, Oxford.

Chapter 5: Death in the brain

1. Whalley, L. (2001) *The Ageing Brain.* Orion Books, London.

2. Hebert, L. E., Scherr, P. A., Bienias, J. L., Bennett, D. A. and Evans, D. A. (2003) Alzheimer disease in the U.S. population: prevalence estimates using the 2000 census. *Archives of Neurology,* **60**, 1119–22.

3. Dementia UK (2007) *A Report Into the Prevalence and Cost of Dementia.* Prepared by the Personal Social Services Research Unit (PSSRU) at the London School of Economics and Institute of Psychiatry at King's College London. The Alzheimer's Society.

4. Evans, D. A., Funkenstein, H. H., Albert, M. S. *et al.* (1989) Prevalence of Alzheimer's disease in a community population of older persons: higher than previously reported. *Journal of the American Medical Association,* **262,** 2552–6.

5. Brown, G. (1999) *The Energy of Life.* HarperCollins, London.

6. Coskun, P. E., Beal, M. F. and Wallace, D. C. (2004) Alzheimer's brains harbor somatic mtDNA control-region mutations that suppress mitochondrial transcription and replication. *Proceedings of the National Academy of Sciences of the USA,* **101,** 10726–31.

7. Bender, A., Krishnan, K. J., Morris, C. M., Taylor, G. A., Reeve, A. K., Perry, R. H., Jaros, E., Hersheson, J. S., Betts, J., Klopstock, T., Taylor, R. W. and Turnbull, D. M. (2006) High levels of mitochondrial DNA deletions in substantia nigra neurons in aging and Parkinson disease. *Nature Genetics,* **38,** 515–17.

8. Esiri, M. M., Matthews, F., Brayne, C. and Ince, P. G. (2001) Pathological correlates of late-onset dementia in a multi-centre, community-based population in England and Wales. *Lancet,* **357,** 169–75.

9. Thal, D. R., Del Tredici, K. and Braak, H. (2004) Neurodegeneration in normal brain aging and disease. *Science of Aging Know-ledge Environment,* **26.**

Chapter 6: Death of self

1. Grand, S. (2000) *Creation: Life and How to Make It.* Weidenfeld & Nicolson, London.

Chapter 7: Cellular death

1. Skulachev, V. P. (1999) Phenoptosis: Programmed death of an organism. *Biochemistry (Moscow)*, **64**, 1418–26.

2. Lewis, K. (1999) Human longevity: an evolutionary approach. *Mechanics of Aging and Development*, **109**, 43–51.

3. Penninx, B. W., Van Tilburg, T., Kriegsman, D. M., Deeg, D.J., Boeke, A. J. and van Eijk, J. T. (1997) *American Journal of Epidemiology*, **146**, 510–19.

Interlude 7: When do we become old?

1. Milton, J. (1667) *Paradise Lost*. Penguin Classics, London.

2. Monois, G. (1989) *History of Old Age, from Antiquity to the Renaissance* (transl. S. H. Tenison). Chicago University Press, Chicago.

Chapter 8: Aging

1. Olshansky, S. J. and Carnes, B. A. (2001) *The Quest for Immortality: Science At the Frontiers of Aging*. WW Norton & Company, New York.

2. Weismann, A. (1889) *Essays upon Heredity and Kindred Biological Problems*. Clarendon Press, Oxford.

3. Medawar, P. B. (1952) *An Unsolved Problem of Biology*. H.K. Lewis, London.

4. Williams, G. C. (1957) Pleiotropy, natural selection and the evolution of senescence. *Evolution*, **11**, 398–411.

5. Kirkwood, T. B. L. (1977) Evolution of aging. *Nature*, **270**, 301–4.

6. Westendorf, R. G. J. (2004) Are we becoming less disposable? *EMBO Reports*, **5**, 1–6.

7. Zafon C. (2003) Aging purpose: another thrifty genotype. *Medical Hypotheses*, **61**, 482–5.

8. Kirkwood, T. B. (2003) The most pressing problem of our age. *British Medical Journal*, **326**, 1297–9.

9. West, M. and Brands, H. W. (2003) *The Immortal Cell: One Scientist's Quest to Solve the Mystery of Human Aging*, Doubleday Books, New York.

10. Austad, Steven N. (1997) *Why We Age*. Wiley, New York.

11. Finkel, T. and Holbrook, N. J. (2000) Oxidants, oxidative stress and the biology of aging. *Nature*, **408**, 239–47.

12. Trifunovic, A. *et al.* (2004) Premature ageing in mice expressing defective mitochondrial DNA polymerase. *Nature*, **429**, 417.

13. DePinho, R. A. (2000) The age of cancer. *Nature*, **408**, 248–54.

14. Olshansky, S. J. *et al.* (2005) A potential decline in life expectancy in the United States in the twenty-first century. *New England Journal of Medicine*, **352**, 1138–45.

15. Valdes, A., Andrew, T., Gardner, J., Kimura, M., Oelsner, E., Cherkas, L., Aviv, A. and Spector, T. (2005) Obesity, cigarette smoking, and telomere length in women. *Lancet*, **366** (9486), 662–4.

16. Rose, M. R., Rauser, C. L., Mueller, L.. D. and Benford, G. (2006) A revolution for aging research. *Biogerontology*, **7**, 269–77.

17. Vaupel, J. W. *et al.* (1998) Biodemographic trajectories of longevity. *Science*, **280**, 855–60.

Interlude 8: Should death be resisted?

1. Bryan Appleyard (2007) How to Live Forever or Die Trying. Simon & Schuster, London.

Chapter 9: Immortality

1. Olshansky, S. J., Carnes, B. A. and Cassel, C. (1990) In search of Methuselah: estimating the upper limits to human longevity. *Science*, **250**, 634–40.

2. Olshansky, S. J. *et al.* (2005) A potential decline in life expectancy in the United States in the 21st century. *New England Journal of Medicine*, **352**, 1138–45.

3. Olshansky, S. J. and Carnes, B. A. (2001) *The Quest for Immortality: Science at the Frontiers of Aging*. WW Norton & Company, New York.

4. Oeppen, J. and Vaupel, J. W. (2002) Demography. Broken limits to life expectancy. *Science*, **296**, 1029–31.

5. Vaupel, J. W. *et al.* (1998) Biodemographic trajectories of longevity. *Science*, **280**, 855–60.

6. Hayflick, L. (1996) *How and Why We Age*. Ballantine Books, New York.

7. Hayflick, L. (2000) The future of ageing. *Nature*, **408**, 267–9.

8. Anderson, R. N. (1999) *US Decennial Life Tables for 1989–91*, Vol. 1, No. 4, pp. 7–8. National Center for Health Statistics, Hyattsville, MD.

9. Thane, P. (2005) *The Long History of Old Age*. Thames & Hudson, London.

10. Austad, S. N. (1997) *Why We Age*. Wiley, New York.

11. Robine, J.-M. and Romieu, I. (1998) *Healthy Active Ageing: Health Expectancies at Age 65 in the Different Parts of the World*. REVES/INSERM, Montpellier.

12. Dementia UK (2007) *A Report Into the Prevalence and Cost of Dementia*. Prepared by the Personal Social Services Research Unit (PSSRU) at the London School of Economics and Institute of Psychiatry at King's College London. The Alzheimer's Society.

13. Yesavage, J. A. *et al.* (2002) Modeling the prevalence and incidence of Alzheimer's disease and mild cognitive impairment. *Journal of Psychiatric Research,* **36,** 281–6.

14. Murray, C. J. L. and Lopez, A. D. (1997) Alternative projections of mortality and disability by cause 1990–2020: Global Burden of Disease Study. *Lancet,* **349,** 1498–504.

15. Hall, S. S. (2005) *Merchants of Immortality: Chasing the Dream of Human Life Extension.* Mariner Books, New York.

16. De Grey A. D. N. J. *et al.* (2002) Time to talk SENS: Critiquing the immutability of human aging. *Annals of the New York Academy of Sciences,* **959,** 452–62.

17. Warner, H. *et al.* (2005) Science fact and the SENS agenda. *EMBO Reports,* **6**(11), 1006–8.

18. Austad, S. N. (1997) *Why We Age.* Wiley, New York.

19. Hughes, J. (2001) The future of death: cryonics and the telos of liberal individualism. *Journal of Evolution and Technology,* **6,** 1–23.

Index

Death is not what it once was. The decline of acute death by infections, starvation, violence and heart attack has allowed people to reach extreme old age, but has ushered in disability, dementia and degenerative disease, with profound consequences for the self and society. The future of death is even more extreme, and constitutes one of the greatest challenges of the 21st century. In chapters echoing Dante's nine circles of hell, Dr Guy Brown explores these vital issues at various levels, from the cell, to the whole body, to society. He reveals that cell death is central to cutting edge biology and medicine, from embryo formation to cancer cures. He tracks the seismic shifts in the causes and character of death that are rocking medicine. And he reveals how technological innovations, such as cloning and electronic interfaces, hint at new modes of survival after death.

Dr Guy Brown heads a research group at the University of Cambridge, UK, working on cell death in the brain, the heart and in cancer. His previous book, *The Energy of Life* (HarperCollins/Simon & Schuster), won the Wellcome Trust Prize for popular science.

macmillan